WHAT THE
NUMBERS
SAY

DERRICK NIEDERMAN
and
DAVID BOYUM

WHAT THE
NUMBERS
SAY A Field Guide

to Mastering

Our Numerical World

BROADWAY BOOKS
NEW YORK

PRINTED IN THE UNITED STATES OF AMERICA

BROADWAY BOOKS and its logo, a letter B bisected
on the diagonal, are trademarks of Random House, Inc.

Visit our website at www.broadwaybooks.com

First edition published July 2003

Library of Congress Cataloging-in-Publication Data

Niederman, Derrick.
What the numbers say : a field guide to mastering
our numerical world / by Derrick Niederman and
David Boyum.—1st ed.
p.   cm.
Includes bibliographical references.
1. Statistics—popular works.   I. Boyum, David.   II. Title.
QA276 .N455 2003
001.4'22—dc21   2002034450

ISBN 0-7679-0998-4

10   9   8   7   6   5   4   3   2   1

To Nancy Camp
and Asmund Boyum,
who got us started on
a lifelong love of numbers

CONTENTS

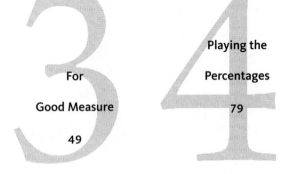
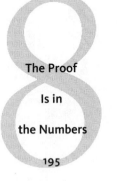

ACKNOWLEDGMENTS

This book was fun to write, but neither one of us would say that it was easy. Fortunately, we had the help of some first-rate editorial minds to inspire us, goad us, and help make the manuscript better at every turn. At the top of the list were Joy Gould Boyum and Peggy Malaspina, who nurtured the book even when it was a mere proposal awaiting the eyes of the publishing community.

Once we started the book in earnest, it became even more apparent that we couldn't go it alone. We received invaluable feedback on our early chapter drafts from the following all-star lineup: Will Carlin, John Chittenden, David Gosse, Mark Kleiman, Carol Loomis, Greg Lyss, Dick Murnane, John Pakutka, Cordelia Tappin, and Charles and Geraldine Van Doren.

We would also like to thank our many friends and family members who endured countless conversations about quantitative reasoning: John Anz, Jon Caulkins, David and Ingrid Ellen, Alexes Hazen, Andy Long, Ted Marmor, Jim and Mimi Niederman, Steve Piltch, Harold Pollack, Peter Reuter, and Rich Zabel.

Our agent, Charles Everitt, represented the book with his accustomed zeal and integrity. We are pleased that he paired us with Gerry Howard at Broadway, who gave us the freedom to create just the book we wanted. To hear him say that the individual chapters were making him feel smarter assured us that we were on the right track, no matter how many zigs and zags we took in reaching our destination. We would also like to thank the other publishing professionals at Broadway who worked hard to prepare our book for the outside world.

Finally, we can think of several individuals who had nothing to do with *What the Numbers Say* but were nonetheless instrumental, in the sense that our project never would have developed without them. That group includes Deborah Hughes Hallett, who hired David to teach quantitative reasoning while he was an undergraduate at Harvard, and the late Irving Ezra Segal, who guided Derrick's graduate thesis at MIT. More recently and perhaps decisively, there was John Allen Paulos, whose inspirational book *Innumeracy* reminded us that there are people out there who love mathematics, fret at its abuse and neglect, and long for a future in which quantitative thinking comes as naturally as A, B, C.

Derrick Niederman
David Boyum
April 2003

The Quantitative Information Age

Whether we realize it or not, quantitative information pervades our professional and personal lives. Every time a doctor reads a patient's chart, a farmer weighs a hog, a businesswoman reviews a budget, or a driver glances at a speedometer, someone is seeking quantitative information. And thanks to computers and the Internet, numbers are spreading faster than deer populations. A few years back, a car shopper found it hard to get much in the way of guidance—a magazine article here, a friend's opinion there. Now, a little web surfing reveals sticker and invoice prices, dealer holdbacks and incentive packages, depreciation estimates, customer satisfaction ratings, reliability statistics, expected insurance costs, crash test results and various accident, injury, and fatality rates, not to mention a wealth of financing data. We may say we live in the "Information Age," but it might be more accurate to say we live in the "Quantitative Information Age."

As technology builds this crescendo of numbers, our ability to make wise decisions, whether at work or at home, increasingly depends on proficiency with quantitative information. Unfortunately, Americans seem much better at producing numbers than making sense of them. It was to underscore this point that cognitive scientist Douglas Hofstadter coined the term "innumeracy," a word brought into the national spotlight by John Allen Paulos's pathbreaking book of the same name.[1] But identifying educational failures is far easier than remedying them, and while countless books, articles, and government reports have diagnosed the widespread ailment of poor quantitative thinking, they have not provided therapy. Or at least not much of it. That's where this book comes in. Our goal is to introduce you to the quantitative concepts, skills, and habits you need for success in work, and success in life.

Sounds dreadful, doesn't it? Yet another remedial math book. Well, if that's what you're expecting, we've got good news. It turns out that formal mathematics is not the best way to teach quantitative thinking, and superior quantitative reasoning is not restricted to those who aced high school math.

Imagine, if you will, the work of an accountant, green eyeshade and all. Few jobs involve greater contact with numbers, and quantitative skills are an obvious prerequisite. If an accountant isn't good with figures, he might not notice a liability that has been excluded from a company's balance sheet, or a depreciation charge that has been miscalculated. Yet how much advanced mathematics is required to assess numbers on financial statements? These numbers get added, subtracted, multiplied, divided, and displayed as fractions, decimals, and percentages. They don't get expressed as Gaussian integers. The last time we looked at a financial statement, we didn't see a ∇, a $\cap$, a $\propto$, or a $\exists$.

Think about other professional occupations—architects, doctors, management consultants, financial planners, marketing executives. Again, large amounts of quantitative information are a feature of many such jobs, and good quantitative thinking is critical to doing the jobs well. But matrix algebra is not required. Nor are high school staples such as quadratic equations, analytic geometry, and imaginary numbers.

This is why high school math teachers fear nothing more than the question, "Why do I need to learn this?" The conventional response, "Because you'll need the skill in your job," is dishonest, because even if the class contains some future engineers, they aren't the ones asking the question. The more honest reply, "Because you'll need it for the SAT," wins points for candor but loses them right back to overt cynicism. Thomas Jefferson's endorsement of mathematics—"The faculties of the mind, like the members of the body, are strengthened and improved by exercise"[2]—is closer to our idea of why the study of math is so essential, but we admit that such an argument will hardly motivate apathetic teenagers.

But if math isn't the key to good quantitative thinking, what in the world is inside this book? Here's a sneak preview. What distinguishes good quantitative thinkers is not their skill with pure mathematics, but rather their approach to quantitative information. Effective quantitative thinkers possess certain attitudes, skills, and habits that they bring to bear whenever they need to make decisions based on numbers. And with rare exception, these attitudes, skills, and habits are not taught in math classes or textbooks. For example, good quantitative thinkers demand empirical evidence instead of conventional wisdom. At the same time, they assume that figures are often wrong or misleading. Good quantitative thinkers don't examine quantitative information until they have a strategy for doing so. They know that some numbers are far more important than others, and they systematically pare data in an effort to find the most revealing figures. They also make lots of rough estimates, scribble on the backs of envelopes, and use arithmetic shortcuts like the Rule of 72. These are the skills we intend to convey.

Now for the most welcome news of all: In showing you the ways of good quantitative thinkers, we are confident that you already know all the math you need. We may jog your memory from time to time, but we're talking about the basics—arith-

metic, percentages, fractions, decimals, square roots, and exponents. If you're still stumped every time you hear someone say,
"That's six of one, a half dozen of the other," then perhaps you do
need a remedial math book. But otherwise you've come to the
right place and we're delighted to have you aboard. Be prepared
to see numbers in a brand new light.

The Ten Habits of Highly Effective Quantitative Thinkers

When Shaquille O'Neal accepted the NBA's Most Valuable Player award in 2000, he quoted Aristotle: "Excellence is not a singular act, but a habit. You are what you repeatedly do."[1] Classicists might quibble with Shaq's translation of *The Nicomachean Ethics*, but in our view he was right on the money. Excellence in anything is the product of practice. That's especially true of quantitative reasoning, which doesn't come naturally to any of us. It seems to be our fate to enter this world with lousy quantitative instincts, as if Adam miscounted the fruit on the tree of knowledge and forced all of us to suffer for his arithmetic sin. Or, to put the same thought in more secular terms, the remote ancestors whose struggle for survival shaped our genetic makeup faced an environment where quantitative skills were not especially important in solving the problem of finding a meal without becoming a meal.

Like the young tourist asking how to get to Carnegie Hall, someone asking the way to good quantitative thinking must be told, "Practice, practice, practice." But what are you practicing? Ideally, habits that foster effective quantitative thinking. Thanks to Stephen Covey's best-selling books (*The 7 Habits of Highly Effective People* and its offshoots), seven has become the canonical number for lists of good habits. However, for those born into a decimal system, seven is an awkward number. Neurologists test for dementia by asking patients to serially subtract 7, starting from 100, precisely because 7 doesn't create simple patterns. Highly effective quantitative thinkers understand the advantages of working with round numbers, particularly powers of ten: quicker calculations, fewer errors, better recall. That, and the examples of Moses and David Letterman, has led us to make a list of ten—or, perhaps, the top ten—habits of effective quantitative thinkers. Some of these habits are intertwined, but that's no problem. Because 10 is a composite number (unlike, for example, 7), we can display our 10 habits in five groups of two. Get ready to practice.

Attitude Is Everything

What makes someone a skilled quantitative thinker, above all, is the ability to effectively use quantitative information when drawing conclusions or making decisions. In that sense, numbers are a form of evidence. And as any experienced judge would tell you, when you're weighing evidence, attitude is all-important.

Habit 1: Only Trust Numbers

Accepting numbers as evidence is more difficult than it sounds. We tend to trust what we experience, and seeing a number on a piece of paper is not much of an experience. Numbers, after all, are just symbolic representations of quantities, quantities that themselves are often more abstract than real. Do you really understand what it means when an air conditioner has an energy efficiency rating of 10.7, other than higher is better? We certainly don't. But if you want to be a good quantitative thinker, you must learn to make decisions on the basis of numerical information, even when that information conflicts with your instincts and perceptions. How? Try to raise your level of trust in careful quantitative analysis, and reduce your confidence in hunches, theories, and casual observation. Sublimate your impulse to leap to conclusions, transforming it into an urge to seek hard data. And keep an open mind when the data don't go your way. In short, only trust numbers.

Enough pop psychology. We'll try to demonstrate why you should only trust numbers.

One of your authors suffered recurrent ear infections as a young child. The infections promptly came to an end when he began taking Sudafed on the advice of a leading otolaryngologist (the fancy name for an ear, nose, and throat doctor). His parents think Sudafed is a wonder drug. He thinks his parents are crediting the rooster for the sunrise.

The otolaryngologist no longer recommends Sudafed for ear infections. He stopped doing so when the numbers came in. Studies that randomly gave Sudafed to some children and sugar pills to other kids found that Sudafed did not affect the frequency or duration of infections.[2] Unlike the otolaryngologist, your author's parents have had a hard time accepting this quantitative evidence. "Look, I know what I saw," they both claim, which is another way of saying that hundreds of unwitnessed outcomes reported by medical researchers are no match for one firsthand observation. Good quantitative thinkers would demand the numbers, but for some reason your author's parents left the Sudafed research to plead its case in the manner of Groucho Marx: "Who are you going to believe, me or your lying eyes?"[3]

So what did put a stop to your author's ear infections right after he started taking Sudafed? Development and coincidence. Most children grow out of ear infections as their middle-ear structures develop and as their immune systems strengthen. So once a kid's susceptibility to ear infections has peaked, any treatment regimen will tend to coincide with reduced incidence of infection. Sudafed, bee pollen, shark cartilage, an extra bowl of Cap'n Crunch every morning—it all works.

Only trusting numbers is especially important when quantitative analysis clashes with people's irrational tendencies—like shortsightedness. When the U.S. military downsized in the early 1990s, thousands of servicemen and women were given the choice of a lump-sum severance payment or an annuity (payments made over a period of years). In comparative terms, the annuity was so favorable that it effectively offered a guaranteed annual return of 17.5 to 19.8 percent to those who declined the lump sum. That's *guaranteed*, as in *risk-free*.

Yet despite receiving pamphlets demonstrating the superiority of the annuity at prevailing money market rates, 92 percent of enlisted personnel and 51 percent of officers chose the lump sum. Granted, a small number may have had pressing financial needs that forced them to pass up what was likely the best investment opportunity they would ever have. However, the rest of those who took the lump sum were simply bad quantitative thinkers. They were shown the numbers, but they could not bring themselves to trust them.

Interestingly, Department of Defense economists, naïvely underestimating the power of immediate gratification to defeat sound quantitative reasoning, had predicted that roughly half of enlisted personnel and almost no officers would pick the lump sum. It's hardly reassuring to learn that the armed forces are teeming with unskilled quantitative thinkers. But look on the bright side. By shortchanging themselves, those who grabbed the up-front cash saved the rest of us taxpayers an estimated $1.7 billion.[4]

By the way, in case our hyperbole has escaped you, we don't literally mean that you should only trust numbers. Of course, other forms of information, as well as verbal, spatial, and other types of reasoning, are also critical to making intelligent judgments. But to counteract the almost universal tendency to undervalue quantitative information and reasoning, *Only Trust Numbers* is a good place to start. Moreover, when you're first learning a new practice, it's often helpful to overemphasize it. So the next time someone makes an unsubstantiated assertion, yell, as if you're Cuba Gooding Jr. in *Jerry Maguire*: "Show me the numbers!"

Habit 2: Never Trust Numbers

Sorry to contradict ourselves so early in the game, but you should never trust numbers. Before we reconcile our apparently inconsistent advice, first let us explain why numbers are not worthy of your trust: It's because numbers can be wrong, are frequently misleading, and all too often have an agenda.

There are a host of reasons why numbers can be wrong. For starters, people lie and cheat. Some quantitative deceit is obvious, as when a teenager says "21" to a bartender, or a used car dealer pitches you a run-down, twelve-year-old coupe showing 38,538 on the odometer. Other deceit is harder to spot, as when a scientist doctors his experimental results, or respondents give inaccurate answers to survey questions.

Numbers are also wrong for more innocent reasons. People are slow to update databases—so, for example, the apple juice marked "Sale $2.49" on the grocery store shelf adds $2.99 to your bill when it's scanned at checkout. People misremember figures, bungle arithmetic, and punch the wrong keys. Computer programs (and some of the processors they run on) have bugs. Scales, radar guns, barometers, and other measuring devices are also imperfect, even when used and calibrated properly.

Even when accurate, numbers can easily mislead. Quantitative data are seductive; they invite us to engage in the risky behavior of reading more into data than is warranted. Between 1980 and 1987 the rate of newly diagnosed breast cancer cases increased by 32 percent in the United States.[5] This was bad news, right? No, it was probably good news. Epidemiologists believe that most or all of the increase was the result of improved detection of breast cancer, a beneficial development. On a lighter note, Wilt Chamberlain made 54 percent of his field goal attempts during his storied career, while Larry Bird sank fewer than half of his tries. Was Wilt the better shooter? No way—Bird took much harder shots.

Still another reason to distrust numbers is that numbers are used to advance agendas. Tell yourself this: Every number I see is generated and presented by people who have an interest in how that number is used or interpreted. You have our permission to suspend this presumption if you're looking at the periodic table—as far as we know, there is no scheming behind the atomic weight of tungsten—but any other suspensions are taken at your own risk. Clearly, a cynical attitude toward quantitative information is appropriate when a Weight Watchers ad sports a woman who has lost fifty pounds. The fine print—"results not typical"—shouldn't surprise you in the least.

What you may not appreciate is the extent to which machinations lie behind numbers. The recommended dosage of Schering-Plough's blockbuster antihistamine, Claritin, is 10 milligrams. We'll forgive you—this time—if you assumed that "recommended" means "most effective." It took more than six years for the FDA to approve Claritin, in part, a General Accounting Office report suggests, because FDA reviewers felt 10-mg doses were not very effective in relieving seasonal allergy symptoms, while 40-mg doses were effective.[6] So why isn't 40 mg the recommended dosage? The answer may lie in the fine print of the Claritin package insert: "In studies with CLARITIN Tablets at doses two to four times higher than the recommended dose of 10 mg, a dose-related increase in the incidence of somnolence was observed." Many believe that Schering-Plough pushed the 10 mg dosage because their marketing plans depended on being able to claim that Claritin was nonsedating. A higher recommended dosage might strike a better balance between allergy relief and adverse reactions such as drowsiness, but it would undermine Schering-Plough's promotional campaign.

You might think that distrusting numbers is easy, but it's not. Distrusting numbers is not the same as disregarding them. Distrusting means taking the time to double-check a bill; disregarding means paying whatever is demanded with alacrity. But distrusting numbers can be more than time-consuming. It can also be awkward, since the person whose numbers you distrust may be a co-worker, friend, or family member. Or a surly cashier who you just know is going to challenge you when you point out that you're owed another 60 cents.

Figuring out the right way to distrust numbers in professional situations is especially difficult. Goals of efficiency and morale require that co-workers have confidence in one another's work. But, as even generals and czars have learned, blind acceptance of numbers is no solution. Shortly before a scheduled trip to the Netherlands, Clinton administration drug czar Gen. Barry McCaffrey called Dutch drug policy "an unmitigated disaster" and backed up that characterization by asserting that "the murder rate in Holland is double that in the United States."[7] McCaffrey relied on figures given to him by his staff, which is no excuse when the numbers don't make sense. Was he truly unaware that America has far more violent crime than Europe? In fact, the Dutch homicide rate is less than a quarter of the American rate.

McCaffrey isn't the only guilty party in this story. Evidently, his staff relied on cross-national crime statistics published by Interpol, data in which the Dutch homicide rate included *attempted* murders. According to *National Journal* columnist Jonathan Rauch, a McCaffrey spokesman tried to pass the buck by saying, "We have said if we are wrong, speak to Interpol—it's not our statistics, it's [their] reporting."[8] Never mind that the relevant Interpol publication prefaces the stats with the following caveat:

> The data gathered in these sets of statistics is not intended to be used as a basis for comparisons between different countries since the statistics cannot take account of the differences which exist between definitions of punishable acts in different national laws, or the diversity of statistical methods, or the changes which may occur during the reference period and affect the data collected.[9]

The bottom line is that if either McCaffrey or his staff had practiced the habit of doubting numbers, this embarrassing incident would have been avoided.

The two habits we have introduced so far—the twin practices of only trusting numbers and never trusting numbers—are not at all contradictory. If anything, they operate hand in hand. Consider how a skilled securities analyst would approach a company's financial statements. He would delve deeply into the numbers, not accepting the upbeat forecasts of the company's executives until he could square them with his own analysis of the firm's financial data. At the same time, he would be skeptical of the company's figures, recognizing that firms stage-manage their reported financial results—generally in allowable ways, but sometimes fraudulently—to gain favor with investors.

Navigational Tools

Columbus would never have been a great explorer if he didn't have navigational techniques to guide him on the open seas. (To be sure, Columbus had some navigational snafus, such as mistaking the Caribbean islands for Asia, but the goal of this book is excellence, not perfection.) Similarly, you'll never be a great quantitative thinker if you can't find your way through quantitative information. Numbers don't come with maps or directions, and without effective navigational tools, you'll either get lost in a sea of numbers or think you've found what you're looking for when you haven't.

Habit 3: Play Jeopardy

In case you're unfamiliar with the TV game show *Jeopardy*, here's the key rule: All answers must be phrased as questions. So if the category is "Convicted Congressmen," and the clue is, "A former inmate of the Federal Corrections Institution in Oxford, Wisconsin, and a former chairman of the Ways and Means Committee," the correct response is, "Who is Dan Rostenkowski?" If the clue is, "An Ohio representative who typically ended his floor speeches with a request to 'beam me up,' " the answer is, "Who is James Traficant?" Get the idea?

With *Jeopardy* in mind, we bring you Habit 3, courtesy of Mark Kleiman, professor of policy studies at UCLA.

> A number only gets to be useful when considered as the answer to a question. To be a good consumer of numbers, the reader must constantly ask himself:
> - To what question is this number (supposed to be) the answer?

- Is it the correct answer to that question?
- Is that the question to which I need the answer?

Already this line of attack makes sense. The only reason anyone recognizes the number 5,280 is that it answers the question, "How many feet are there in a mile?" Now let's see how the habit works in practice. You are shopping for new wheels and check out the sticker of a car you're interested in. "Manufacturer's Suggested Retail Price $32,320," it says. Time to play *Jeopardy*.

To what question is this number (supposed to be) the answer? Normally, the dollar figure on a price tag indicates what you have to pay to buy an item. When you see "$1.79 %10" on a gasoline pump, the price of a gallon of gas is $1.799. So although $32,320 is modified by the telltale word "suggested," the sticker implies that $32,320 is the answer to the question, "What is the price of this car?"

Is it the correct answer to that question? No. For one thing, most automakers exclude some kind of destination and handling charge from the Manufacturer's Suggested Retail Price (MSRP). More important, the price at which a given new car can be had is a reflection of supply and demand (and negotiating savvy), not what the manufacturer suggests. At the time this was written, we compared the prices a major buying service offered on a pair of 2001 vehicles, the Acura MDX and the Oldsmobile Aurora. The MDX was selling like hotcakes. The Aurora was selling like—well, like an Oldsmobile, a brand of such declining popularity that General Motors had already decided to phase it out. The MSRP (including destination) on the MDX was $34,850; the buying service price was $35,850. The MSRP on the Aurora model we selected for comparison was $35,464; the buying service price was $30,824. The MDX was selling for $1,000 above list, the Aurora for $4,640 below. In other words, although the MDX had a lower MSRP than the Aurora, the price was really $5,000 more.

Is that the question to which I need the answer? No. The question to which you need the answer is: How much it will cost to own the car? And purchase price is not the answer to that question. Ownership costs include depreciation, financing (the interest paid on any money borrowed to buy the car plus the interest not earned on any cash payment), insurance, registration fees, taxes, fuel, maintenance, repairs, and parking. We suspect that few car shoppers take the time to estimate all of these costs, and as a consequence, few realize how quickly the costs add up, and also just how inadequate price is as a measure of ownership costs.

Intellichoice is a California outfit that publishes the *The Complete Car Cost Guide*, an annual volume that should be required reading for economy-minded car shoppers. Consider the typical five-year ownership cost Intellichoice projected for two 2001 four-door sedans, the compact Kia Sephia and the midsize Honda Accord DX. The Sephia listed for $10,595, and Intellichoice estimated it could be bought for $9,824; the Accord had a $15,400 sticker and an assumed purchase price of $13,977.

	Kia Sephia	Honda Accord DX
Depreciation	$7,619	$5,730
Financing	$2,471	$3,374
Insurance	$9,752	$8,390
State Fees	$493	$678
Fuel	$4,006	$4,163
Maintenance	$1,254	$1,022
Repairs	$701	$565
5-Year Ownership Cost	$26,296	$23,922

Look what playing *Jeopardy* reveals. Asking and answering the question of what it would cost to own these cars provides a wealth of information not captured in the sticker prices. Did you know that insurance is usually the biggest cost of owning an inexpensive car? (Depreciation is the largest cost of expensive cars.) Would you have guessed that over five years it regularly costs more than $25,000 to own a Kia Sephia, a car that sells for less than ten grand? Did you know that although it "costs" around 40 percent more to buy a Honda Accord than a Kia Sephia, an Accord is typically less expensive, all things considered?

If you happen to own a Honda Accord, here's a drill you can use to practice Habit 3. The next time you find yourself alongside a Kia Sephia, roll down your window and tell the other driver, "Hey, I was thinking of buying this exact car, but I couldn't afford it."

Habit 4: Live by Pareto's Law

In the late nineteenth century, Vilfredo Pareto, an Italian economist and sociologist, observed remarkable similarity in the distribution of income and wealth in different countries. Everywhere he looked, these monetary measures were unequally distributed, and the patterns of distribution appeared nearly identical. This was not a chance event, in Pareto's view. He posited that the distribution of income and wealth in every society follows a specific mathematical function, a function now termed a Pareto distribution.

There's no need to concern ourselves with the mathematical details of Pareto's work. What's important for our purposes is to understand the gist of a Pareto distribution, and that can be easily done without recourse to equations. In a Pareto distribution, the frequency of a quantity is inversely related to its size: small quantities appear frequently, while large quantities are rare. If we're looking at the distribution of wealth, as Pareto did, we'll find a handful of billionaires, a modest number of millionaires, and a whole lot of folks who are flat broke.

Subsequent to Pareto, others noticed that many things besides income and wealth seemed to fit a Pareto distribution. There are a few big cities, but many small towns; a few major earthquakes, but many minor temblors. People also began to focus on a particular implication of a Pareto distribution—that a small proportion of items constitutes a large share of the cumulative distribution. For example, the most active 10 percent of burglars are responsible for the majority of housebreaking.[10] In the movie industry, a few blockbuster films dominate studio revenues.[11] A few wardrobe items account for most visits to a clothes closet (creating the annoyance that closets are stuffed with clothes that are rarely worn). This type of imbalance is often called "Pareto's Law," a term generally traced to quality control expert J. M. Juran, who in 1950 referenced Pareto in describing how relatively few types of manufacturing defects accounted for the bulk of quality losses.[12]

All good quantitative thinkers are Paretians. That's because Paretian thinking is crucial to understanding many numbers. For instance, you've no doubt heard that about half of all marriages in America will fail. That statistic is often interpreted to mean that every other person who ties the knot will suffer a divorce. But Elizabeth Taylor's Law (the marital version of Pareto's Law) reminds us that a small fraction of the population accounts for a disproportionate share of divorces, which leads to the conclusion that the majority of people who marry never divorce. Suppose we observed the marital careers of five individuals. One married and divorced three times, another wed twice and split up once, and each of the remaining three had a single, lasting marriage. Overall, four of the eight marriages (50 percent) ended in divorce. But three of the five people (60 percent) had lifelong unions.

Juran also referred to Pareto's Law as the rule of the "vital few and trivial many."* The latter phrase proved less catchy, but it highlights what Juran considered the key lesson of Pareto's Law: Pay attention to the vital few. It is estimated that 10 percent of cars on the road generate more than half of auto-related pollution.[13] Pareto's Law suggests that pollution can be reduced more cheaply by identifying these cars and getting them repaired or junked—as opposed to requiring stricter emissions standards on new vehicles, an approach that doesn't target the heaviest polluters.[14] In fact, if pollution regulations raise the cost of new cars, many car owners will hold on to their older (and more polluting) vehicles for longer than they otherwise would.

There's a counterargument to Pareto's Law, and we've all heard it. "Little things add up." Business lore holds that American Airlines saved $100,000 by eliminating one olive from every salad served in first class.[15] But for American Airlines, $100,000 is peanuts. When Pareto distributions are involved, little things usually add up to little numbers. Big things—such as what American spends on jet fuel—are often big all by themselves. That's not to say that the little-things adage is useless, but its main value lies in fostering an economizing spirit that brings savings on big-ticket items.

Pareto's Law teaches us that numbers have to be prioritized. Whether we're looking at a spreadsheet of financial data, a page of sports stats, or a nutritional label on a cereal box, some numbers are invariably more important than others. A useful technique when looking at a group of numbers is to ask yourself which number, or set of numbers, is the most important. In our experience, asking this question often leads to the discovery that we've been looking at numbers aimlessly, without any game plan. Asking what the most important number is forces us to figure out

---

*The most common synonyms are the "Pareto Principle" and the "80/20 Principle." We prefer "Pareto's Law" to these alternatives, since a law suggests a natural regularity while a principle suggests a maxim of choice.

why we looked at the numbers in the first place, what we are trying to learn from them, and which number or numbers can (and
can't) tell us what we want to know.

Pareto's Law is just as important at home as it is at work. According to surveys, working Americans save much less for retirement every year than they think they should. The annual gap
between desired and actual saving is, on average, about 10 percent
of household income. This is why family discussions so often include such lines as, "You paid WHAT to get your hair colored?"
or, "Just because your best friend has titanium golf clubs doesn't
mean you need them."

Hairstyling and golf equipment are not trivial expenses for most
people, but unless you have little income, or a passion for changing
your hair color or your golf clubs, they don't come close to 10 percent of income. Personal spending, like corporate spending, is gov-
erned by Pareto's Law. While writing this chapter, one of us looked
at a year's worth of his credit card charges, a frightening research
exercise. Five percent of charges accounted for 46 percent of total
expenses. Had your author included his check writing, by which he
pays for most major outlays, 5 percent of items would have easily
accounted for more than half of total expenses.

Tightening your belt by 10 percent is difficult. If you try to
buck Pareto's Law, it's that much harder. According to data from
the federal government's Consumer Expenditure Survey, 32.4 percent of household expenditures go for housing and 19.5 percent
for transportation. That's a total of 51.9 percent, which means
that if you don't cut back on housing or car expenses, you have to
rein in everything else by more than 20 percent. More than that
actually, since outlays for health care, education, and pension and
Social Security contributions account for 14.9 percent of expenditures, and you don't want to pare those. So you'd really have to reduce your spending on everything else by 30 percent to meet the

10 percent target. Let's face up to Pareto's Law: Most Americans need to take a hard look at their housing and car expenses.

Illuminating Numbers

When skilled quantitative thinkers daydream, they imagine a world where numbers are always accompanied by a critical analysis of their significance. The fantasy takes them away from the reality of planet Earth, where trying to comprehend quantitative information is a never-ending struggle. Fortunately, good quantitative thinkers have developed a couple of habits that make it easier to understand numbers.

Habit 5: Play 20 Questions

Your game playing shouldn't end with *Jeopardy*. Recall the old car-ride game "20 Questions," in which you tried to guess a person, place, or thing by asking fewer than twenty yes-or-no questions? Well, if you come across a number you don't understand and you want to flush out its true story, you will have to play some form of this game.

The process of asking follow-up questions has the backing of history. In 1967, Secretary of Defense Robert McNamara asked for classified answers to questions about the Vietnam War. In keeping with our first habit—*Only Trust Numbers*—he began with quantitative inquiries: "Are we lying about the number killed in action? Are our data on pacification accurate?" He then moved on to what the data might show: "Are the services lying to the civilian leaders? Are the civilian leaders lying to the American people?" And finally came the inevitable question, "Can we win this war?"[16] The answers to these and other questions eventually reached the public in the form of the *Pentagon Papers*.

Note that the benefits of a string of questions are not always understood when the process begins. That's why experienced "20 Questions" players begin with broad queries and get more specific as the game proceeds. (In that spirit, one of the less effective opening salvos of four-year-old Spencer Boyum was "Is it a buffalo?") The process is inherently inefficient, but it's amazing how a series of questions can home in on an answer that at first could have been anything whatsoever.

We ask questions in our everyday lives as a means of obtaining clarity, even if we have to jettison the yes-or-no rule. If you need to pick up a friend at the airport, you begin by asking when the flight is scheduled to land. But you're not done. At some point you will surely have asked your friend, "What airline are you flying on?" Later on, you might have checked the airline's website to answer the question, "Is flight 521 on time?" And before setting out, you might have asked yourself, "What's the traffic like at that time of day?" These are the sorts of questions we routinely ask in an effort to save time and trouble.

Of course, we're not expecting anyone to be particularly impressed by our suggestion that asking questions is a good thing. Last we looked, there wasn't anyone out there claiming that asking questions is a *bad* idea. But if asking questions about numbers is such an obvious approach, why don't people do it more frequently? The answer is pretty obvious: We don't ask questions because we're afraid of looking stupid. But maybe, just maybe, that fear is overblown. An amusing yet hopeful episode in this direction took place in 1988, when one of your authors was in an editorial session with famed mutual fund manager Peter Lynch, working on Lynch's first book, *One Up on Wall Street*. At one point a young technology analyst popped in and excitedly announced that some monthly index had just come in at 87.5, or something like that. Lynch looked up and asked, "I take it that's a good number?" The analyst nodded his head vigorously and moved on.

Should 87.5 have meant something all by itself? In other words, did Lynch look stupid for not understanding a number that the junior analyst felt sure he would understand? Not at all. When you make a simple inquiry (and surely there's no simpler inquiry than "Is that good?"), you'll be surprised how often you get a straightforward answer, not the feared rolling of the eyebrows to suggest how stupid you are. That's true even if you don't share Lynch's advantage of being the top dog.

There are stupid questions in this world, of course. For the most part, stupid questions arise when we feel under pressure to ask *something*, either for the sake of conversation or participation. Suburban parents spend years asking each other what schools their kids attend, only proving that they weren't listening in the first place. "Tell me again, where is Jessica in college?" And the more time you have to prepare your questions, the less slack you'll be cut. Ken Fisher, author of *Super Stocks*, says that when he asks a company's CFO about a particular figure only to hear that it's in the latest 10-Q filing, he kicks himself for wasting the time of a valuable contact.

We hope it's comforting to hear that asking straightforward questions to correct near-term misunderstandings needn't get you into trouble. The flip side, however, is that we do get into trouble when we assume we know what a number represents and therefore don't make the inquiries that would straighten us out. To show how this works, we'll try to answer the simplest of questions: How many speeds are there on a 27-speed bicycle?

Obviously we're already applying our second habit: *Never Trust Numbers.* But the Grant's Tomb answer of 27 looks suspiciously high. Four to six speeds are sufficient for a car; why would a bicycle have 27? True, a piano has 88 keys, but the difference between adjacent notes is unmistakable. Could the average cyclist really distinguish among, or make effective use of, 27 gradations?

Our game of 20 Questions begins by asking how those 27 speeds are differentiated. The answer demands a brief review of how a bicycle is powered, but basically all we're talking about is a chain connecting two sets of gears. The bigger of these two sets, located between the two pedals, are the chainrings—a set of three concentric metal rings, each ring having a different size and a different number of teeth. For example, the three rings on a Bianchi Eros, to pick a popular Italian road bike, have 30, 42, and 52 teeth, going from left to right as the seated rider would view them. At any given moment, the chain connects one of those three rings to a cog in a so-called "cassette" in the rear wheel. (A lot of jargon, but that's the way it goes.) As was the case with the chainrings, these cogs are distinguished by the number of teeth: The values for the cassette on the Bianchi bike, this time going from right to left, are 13, 14, 15, 16, 17, 19, 21, 23, and 26. *Jeopardy* players will note that there are a total of nine cogs: Nine times three equals 27, so "27" is the answer to the question, "What is the total number of combinations among nine cogs and three chainrings?" But is that the question to which we need an answer? Does each combination actually represent a different speed?

At this point we have to ask another question: How in the world are those alleged 27 speeds quantified? The answer is that for each potential chainring/cog combination, we calculate the distance the bike would travel if the crankarms (the levers to which the pedals are attached) were given a full revolution. Better yet, the manufacturers have already done the work for us. Below are the relevant numbers for the Campagnolo components found on the Bianchi bike, courtesy of Campagnolo's website:

DISTANCE PER CRANKARM REVOLUTION (IN METERS)

Number of	Number of Chainring Teeth		
Cog Teeth	30	42	52
13	4.84	6.77	8.38
14	4.49	6.29	7.79
15	4.19	5.87	7.27
16	3.93	5.50	6.81
17	3.70	5.18	6.41
19	3.31	4.63	5.74
21	2.99	4.19	5.19
23	2.73	3.83	4.74
26	2.42	3.39	4.19

Such numbers are usually called "gear-inch ratios," which hark back to the days when bicycles had an oversized front wheel. There are alternative calculations and presentations of these numbers, but in all cases they are a numeric proxy for the "speed" of any given combination. But wait a minute. The combinations 30/15, 42/21, and 52/26 all produce the same gear-inch ratio: 4.19. So much for 27 speeds.

Looking at the numbers again, we see that the entire middle column appears to contribute nothing to a bike's overall range. But the overlap is essential; without the middle chainring, traveling between the extremes would require an inordinate number of shifts. And, as a practical matter, combinations at the top of the first column and the bottom of the third column are all but ruled out in the cycling world, because those settings would place the chain on a diagonal, causing needless wear and tear.

Clearly then, there's no definitive answer to the question of how many speeds there are on a 27-speed bike. But it would be poor form to completely dodge our own question, so here's a stab at an answer. A typical recreational cyclist would primarily use the 42-tooth chainring, shifting to the 30-tooth ring for steep climbs, and to the 52-tooth ring at high speeds. The 42-tooth ring provides 9 gear-inch ratios. If you look at the table, the 30-tooth ring provides an additional 3 gear-inch ratios that are significantly lower than any of the 42-tooth ratios, while the 52-tooth ring provides 3 speeds that are significantly higher than any of the 42-tooth speeds. That's a total of 15 gear-inch ratios. So that's our answer: A 27-speed bike has 15 speeds. And we didn't even need all of our 20 questions.

Habit 6: Build Models

Superior quantitative thinkers are habitual model builders. Don't worry, we're not going to tell you to build model airplanes or ships—glue fumes do not improve quantitative reasoning. What we mean is that superior quantitative thinkers constantly create numerical models that simulate the real-world situations they are trying to understand.

For a brief case study, let's try to make some sense out of figure skating. If you've watched figure skating on television, you know that the men's and women's individual competition consists of two parts: the short program and the free skate. (School figures, which tested a skater's ability to trace out specific patterns etched on the ice, were eliminated from international competitions in 1990.) Announcers will remind you that the short program accounts for one-third of the total score while the free skate, or long program, accounts for the remaining two-thirds. The question we will try to answer is "Is that really true?"

In pursuit of our goal, we had better ask the fundamental

question, "How does the scoring system work?" The answer is
that for each of the two segments of the competition, skaters earn
so-called "ordinal" placements based on the votes of nine individual judges (with some fine print thrown in to break ties). The
skater with the highest marks earns a "1," the second-place
finisher gets a "2," and so on. The two ordinal scores are then
summed together to give a final score, with one extra wrinkle:
The score for the short program is multiplied by $\frac{1}{2}$, while the ordinal for the free skate remains unchanged. The adjustments give
rise to the name "factored placements," and they appear to account for the $\frac{1}{3}$ and $\frac{2}{3}$ weightings upon which the system is
built. A skater who wins both segments of the competition will
win the whole event with a score of 1.5, the lowest score possible. Otherwise, the lowest combined score wins. What if there's
a tie? Good question. In the event of a tie, the title goes to the
person with the best free skate.

We're ready for the all-important question: Does this system
actually work the way it's supposed to? The answer is a resounding
"no."

To get to the bottom of this one, we build the simplest model
imaginable. Suppose that Sonja Henie, Peggy Fleming, and Kata-
rina Witt were the only three competitors in the Olympics. Here
is one possible scenario:

Skater	Short Program (S)	Long Program (L)	Final Score ($\frac{1}{2} \times$ S + L)
Henie	3	1	2.5
Fleming	1	2	2.5
Witt	2	3	4

As you can see, Sonja Henie and Peggy Fleming earned the same total score of 2.5, but since a tie goes to the winner of the free skate, Henie wins the whole thing. You see the problem? Henie did as poorly as she could have in the short program and won anyway. In a three-skater field, the effective weighting of the free skate is *100 percent*. That's not necessarily the case in a larger field, but there will be many occasions in which the top three skaters are so ahead of the rest that no one else matters.

Of course, as long as we're talking about the theoretical weaknesses of an ordinal system, we might as well include a real-life glitch. The 2002 Winter Olympics are best known for the judging scandal surrounding the pairs competition, but the women's event had an absurdity of its own. Building on our model above, these were the actual finishes of the top four skaters:

Skater	Short Program (S)	Long Program (L)	Final Score ($\frac{1}{2} \times S + L$)
Sarah Hughes	4	1	3
Irina Slutskaya	2	2	3
Michelle Kwan	1	3	3.5
Sasha Cohen	3	4	5.5

Hughes and Slutskaya finished with identical 3.0 scores, but Hughes got the gold because of the free-skate tiebreaker. What the table doesn't reveal is that Slutskaya was the final skater that night. This might not seem to matter, but what if she had finished third in the free skate rather than second? In that case, Kwan would have walked away with the gold, as follows:

Skater	Short Program (S)	Long Program (L)	Final Score ($\frac{1}{2} \times$ S + L)
Michelle Kwan	1	2	2.5
Sarah Hughes	4	1	3
Irina Slutskaya	2	3	4
Sasha Cohen	3	4	5.5

In other words, the order between Hughes and Kwan in the final standings was ultimately determined by the free-skate performance of . . . Slutskaya. In academic circles, this situation might be seen as a violation of "the independence of irrelevant alternatives," an axiom put forth by none other than John Nash, the protagonist of *A Beautiful Mind*. But we'll just call it a violation of common sense.

Models can be used to analyze all sorts of quantitative circumstances. The idea is to create simplified examples using easy round numbers. For instance, when we looked at Elizabeth Taylor's Law, we imagined five people who among them had eight marriages and four divorces. The numbers quickly and easily showed that a divorce rate of 50 percent was consistent with 60 percent of married people never divorcing.

Social scientists have long known that when the white population in a racially integrated neighborhood is declining, the explanation needn't be "white flight." A model builder would reach the same conclusion by inventing Zebraville, a neighborhood of exactly 500 whites and 500 blacks. The advantage of a simple model with round numbers is that it is easy to play out different scenarios. Say that 30 whites and 70 blacks leave Zebraville, and into their vacated apartments and houses move 20 whites and 80 blacks. Zebraville's racial mix has changed. With 490 whites and 510 blacks, the neighborhood is now 49 percent white and 51 percent black. But it can't be right to point a finger at "white flight" in accounting for the decline in the white population. Whites were less likely than blacks to leave the neighborhood. There was more "black flight" than "white flight."

The model reminds us that there are four factors that determine Zebraville's future racial mix: the number of whites that move out, the number of whites that move in, the number of blacks that exit the neighborhood, and the number of blacks that enter. Note that the "white flight" assumption only looks at one of these four factors. According to Ingrid Gould Ellen, author of *Sharing America's Neighborhoods: Prospects for Stable Racial Integration* (and sister of one of your authors), when the white population in integrated neighborhoods declines, the principal cause is usually "white avoidance," not white flight.[17]

This distinction is more than just semantic. For one thing, it tells us that whites are more satisfied living in integrated neighborhoods than the erroneous "white flight" assumption suggests, an encouraging fact that could reduce white avoidance of integrated neighborhoods if it were more widely known. Also, neighborhoods fearful of racial change typically focus preventive efforts on exit decisions rather than entrance decisions, taking steps such as banning "For Sale" signs and otherwise making life difficult for sellers and their realtors. Instead, it would be wiser for integrated neighborhoods to spend their energy marketing themselves to the outside world.

Uncertainty

As if we haven't given you enough reasons to be careful with numbers, here's another: uncertainty. If a blood test finds that your total cholesterol is 200 mg/dl, you can't know if your cholesterol level was exactly 200, or if your cholesterol was at a representative level at the time of the test, or precisely what a measurement of 200 implies about your coronary artery disease risk. If such lack of clarity discomforts you, then you need to work on Habits 7 and 8.

Habit 7: Play the Odds

When we get around to building a Hall of Fame for highly effective quantitative thinkers, Robert Rubin will be one of the first inductees. Throughout his career, which includes stints as co-chairman of Goldman Sachs and secretary of the Treasury, Rubin has been considered one of the world's coolest decision-makers. On Wall Street, Rubin was a legendary risk arbitrageur, a line of work that entails making huge bets on corporate mergers and acquisitions. In Washington, where the stakes involved the solvency of nations, Rubin was said to be the most influential Treasury secretary since Alexander Hamilton.

What can this future Hall of Famer teach us? We can learn something from his reputation. At Goldman Sachs and in the Clinton administration, Rubin was known for taking a highly analytical approach to problems, carefully exploring and weighing every aspect of potential decisions. But we can learn even more from Rubin himself. Speaking at the University of Pennsylvania commencement in 1999, Rubin offered this summary of his M.O.:

> As I think back over the years, I have been guided by four principles for decision making. First, the only certainty is that there is no certainty. Second, every decision, as a consequence, is a matter of weighing probabilities. Third, despite uncertainty we must decide and we must act. And lastly, we need to judge decisions not only on the results, but on how they were made.[18]

Stop nodding in agreement. If you are at all typical, you have a different M.O. Most people are in denial about uncertainty. They assume they're lucky, and that the unpredictable can be reliably forecast. This keeps business brisk for palm readers, psychics, and stockbrokers, but it's a terrible way to deal with uncertainty.

Bob Rubin is not normal. He doesn't deny uncertainty; he embraces it. "If there are no absolutes," Rubin says cheerfully, as if haphazardness comforts him, "then all decisions become matters of judging the probability of different outcomes, and the costs and benefits of each. Then, on that basis, you can make a good decision." It should be obvious that an honest assessment of uncertainty leads to better decisions, but the benefits of Rubin's approach go beyond that. For starters, although it may seem contradictory, embracing uncertainty reduces risk while denial increases it. Just listen to Rubin again:

> I remember once, many years ago, when a securities trader at another firm told me he had purchased a large block of stock. He did this because he was sure—absolutely certain—a particular set of events would occur. I looked, and I agreed that there were no evident roadblocks. He, with his absolute belief, took a very, very large position. I, highly optimistic but recognizing uncertainty, took a large position. Something totally unexpected happened. The projected events did not occur. I caused my firm to lose a lot of money, but not more than it could absorb. He lost an amount way beyond reason—and his job.

As Mark Twain reportedly said (but according to scholars probably didn't), "The art of prophecy is very difficult, especially with respect to the future."[19]

Another benefit of acknowledging uncertainty is that it keeps you honest. "A healthy respect for uncertainty and focus on probability," Rubin explains, "drives you never to be satisfied with your conclusions. It keeps you moving forward to seek out more information, to question conventional thinking and to continually refine your judgments. And understanding that difference between certainty and likelihood can make all the difference. It might even save your job." Renowned financier George Soros, who calls himself an "insecurity analyst," makes the same point in different terms. "I recognize that I may be wrong. This makes me insecure. My sense of insecurity keeps me alert, always ready to correct my errors."[20]

Lastly, an appreciation of uncertainty can help you deal more effectively and comfortably with bad outcomes. An unavoidable consequence of uncertainty is that good decisions often have bad outcomes. After carefully weighing the relevant probabilities, a forty-two-year-old pregnant woman chooses to have amniocentesis, concluding that it is worth accepting the miscarriage risk posed by the procedure in order to definitively detect or rule out Down's syndrome. The sample of amniotic fluid indicates a normal fetus, but sadly the woman miscarries. Did she make a bad decision? No. Sometimes chance is cruel.

Adopting Rubin's oddsmaker perspective is difficult, especially for control freaks. The reality is that we can't control outcomes; the most we can do is influence the probability of certain outcomes. Exercise regularly and you're likely to live longer—but you might drop dead on a treadmill. No matter how hard we try to govern things, Lady Luck will still have her say. Or, as MIT psychologist Steven Pinker eloquently puts it, "Life is not chess but backgammon, with a throw of the dice at every turn."[21]

A few years ago, one of us worked on a consulting project for a Fortune 500 manufacturing firm. The chief information officer (CIO), new to the job and the company, wanted an assessment of the plans for a major computer project that had been approved prior to his arrival. The company planned to install in one of its largest divisions an enterprise-wide system that would integrate financial, accounting, manufacturing and logistics, engineering, and sales and distribution systems.

The project was expected to take six years to fully implement. The company had engaged a well-known consulting firm to draft an implementation plan. The attention to detail was truly impressive. Every step was carefully and clearly laid out, week-by-week, for six years. At first glance the plan inspired confidence.

But the blueprint had an Achilles' heel: It denied uncertainty. There were countless reasons why the project might not develop according to plan. It was technologically ambitious. It required substantial and potentially divisive changes to organizational structures and operational processes. Over six years, considerable changes to the company's business, both internal and external, were likely. Already the firm was entertaining the ideas of moving certain manufacturing operations overseas and merging the division in question with others. Yet the project scheme left no room for adapting to developments inconsistent with its game plan. The only mechanism it offered for midcourse corrections was to cancel parts of the project that weren't meeting goals. Your author's team argued that the implementation plan offered a false sense of security, and that by overly restricting options it actually increased project risk. The team recommended that the implementation plan be revised to explicitly acknowledge the uncertainty of future developments and lay the groundwork for modifying the project on the fly.

Unfortunately—for both your author and the firm's shareholders—the company's senior managers bore little intellectual resemblance to Bob Rubin. Such an "indeterminate" and "open-ended" approach would be "too risky," they complained. Your author's team was fired. A follow-up call to the CIO a year later confirmed the expected. The project had already run aground.

We'll give Rubin (this time speaking to Harvard Business

School graduates) the last word. "If you can internalize a probabilistic mindset and live by that discipline, you'll be well prepared for the uncertainties and complexities of life and your undertakings. Alternatively, if you think in terms of absolutes and see things in black and white, sooner or later you'll fall over a cliff, most likely sooner."[22]

Habit 8: Know What You Know and Don't Know

Rubin's comments highlight the importance of a probabilistic mindset when making decisions. But thinking like an oddsmaker is also critical when assessing information. There's uncertainty in almost every piece of information. Doctors repeat tests because they know results vary. Companies take write-offs or otherwise revise financial results because originally reported figures prove inaccurate.

Most data should be viewed probabilistically, as one of many possible outcomes, some being more likely than others. If, for example, your flight is scheduled to arrive at Chicago's O'Hare airport at 5:00 P.M., that doesn't mean you're going to disembark in the Windy City at five o'clock, although you might. With O'Hare, anything between, say, 4:00 P.M. and next Tuesday is possible. 5:15 might be a better bet than 5:00, but 5:00 is more probable than 7:00, which is in turn more likely than 4:00.

Data are always imperfect or incomplete. A consequence is that there is often more than one possible account of the data. In such cases, alternative explanations should be viewed probabilistically. Amazingly, this point appears to have been lost on Sir R. A. Fisher, arguably the most important statistician of the twentieth century. In 1958, Fisher published several articles criticizing studies that asserted a causal link between cigarette smoking and lung cancer. Fisher identified several methodological weaknesses in the research, and suggested that the observed correlation between smoking and lung cancer might be explained by some people having a genetic disposition to both smoking and lung cancer. (The idea wasn't entirely speculative; Fisher showed that identical twins were more alike in their smoking behavior than were fraternal twins.) Fisher was right that the smoking studies were flawed; but he was wrong to suggest, as he did, that the data were so inconclusive that any resulting judgments about smoking and lung cancer were unwarranted.

It was conceivable that tobacco wasn't carcinogenic. But any reasonable reading of the data available at the time would have led to the conclusion that it was highly likely that smoking was a major contributor to the rising incidence of lung cancer. Lung cancer was almost nonexistent prior to widespread smoking. Toxicological research demonstrated that tobacco smoke had carcinogenic effects on different types of animal tissue. Numerous studies in different countries showed that lung cancer rates were many times higher among smokers than nonsmokers. Moreover, studies that looked at how much people smoked found that heavier smokers had higher rates of lung cancer than lighter smokers. Such a "dose response" is the key indicator of an effect in pharmacology.[23]

The Fisher story brings one of us back to a Harvard classroom. When your author dismissed some data as meaningless, Professor Mark Moore challenged him. "It's hard to believe the data contain *no* information," Moore suggested. "Surely there is *some* information. What is it?" Á la Rubin, Moore often asked students to assign probabilities to different conjectures, forcing them to both recognize the uncertainty of their knowledge and to carefully assess what they knew. "If you had to bet," Moore would have asked Fisher, "where would you put your money? Would you bet that smoking contributes to lung cancer, or would you wager that it doesn't?"

To deal with the uncertainty in our knowledge, it is useful to make a practice of thoroughly taking stock of what you know and what you don't know. This habit reduces the chance that you will overestimate or underestimate your knowledge. It also guides your learning. If you don't know what you don't know, then you also don't know what information would improve your knowledge. And knowing what you know and don't know is a prerequisite to making the kind of probabilistic judgments that Rubin and Moore recommend.

Put yourself in the shoes of a high school student who is considering where she should apply to college. Browsing the Bryn Mawr website, she finds a page titled, "Why Choose a College for Women."[24] "Studies show that women's college graduates are more likely to earn an advanced degree," it says. And, "Studies show that women's college graduates are more successful in their careers and hold higher positions than women who graduate from coeducational institutions." What should this college shopper make of this information?

Does she know if the statements are accurate? No. It seems unlikely that a leading college would falsify such information, but it's worth checking, and some more surfing confirms that Bryn Mawr has not lied. A bigger question is: Does she know why graduates of women's colleges are relatively more successful? Again, no, but she can entertain some hypotheses. Bryn Mawr presumably wants prospective applicants to believe that graduates of women's colleges do well because their schools do a better job of educating and training women than do coeducational schools. But it is also plausible that the higher achievement of women's college graduates is partly or completely attributable to the graduates themselves rather than their schools. Perhaps women who attend Bryn Mawr, Barnard, and other women's colleges are, on average, smarter, more disciplined, and more career-oriented than women who enroll in coed schools. Wasn't Hillary Clinton bright and ambitious *before* she got to Wellesley?

There is more that our high school student doesn't know. Even if it is true that women will, on average, be more successful if they attend women's colleges as opposed to comparably selective coed schools, it does not follow that all women's colleges are better on that score. Which women's colleges are educationally most effective? (Just to clarify, the issue here is not how Wellesley compares with Sweet Briar, but how Wellesley and Sweet Briar compare with their respective coed peers.) And it is surely the case that whatever the averages show, some women will fare better in a single-sex environment while others will achieve more in a coed setting. Which type of woman is our high school student?

None of these questions can be answered with much certainty. But if our high school student does a little research and continues to ask herself what she knows and doesn't know, she will improve her chances of coming up with good answers. Which will improve her chances of making a wise college selection.

Estimation

Do you dress differently on a 63-degree day than on a 65-degree day? If someone tells you that the Dow Jones Industrial Average rose 72 points on a particular day, do you feel betrayed when you find out that the actual gain was only 71.78? The point is that the world isn't as precise as we are sometimes led to believe. Even in the twenty-first century, close can be good enough, and our final set of habits is geared to acknowledge and exploit this simple fact of life.

Habit 9: Go Figure

If the following habit doesn't make us look old-fashioned, nothing will, because we're about to extol the virtues of making calculations by hand. We're painfully aware that antediluvian skills such as rounding and estimating are a tough sell in an age when electronics can create decimal-point precision. But we're also mindful of the exasperation of friend and high school physics teacher Ronald Newburgh, who once gave his students a couple of equations and a few data points, and asked them to calculate the weight of a particularly heavy object. One student proudly came back with an answer of 3,000 kilograms, which, as Newburgh noted, might have been a reasonable stab had the object in question been a Mercedes. Unfortunately, the class had been asked to estimate the weight of *Earth*.

Machines cannot substitute for the skill of being able to size up a number and ask "Does that look right?" Nor can they guarantee that we are plugging in the right numbers or that we are performing the right calculation in the first place. The expression "garbage-in, garbage-out" is strictly a modern-day coinage. It's tempting to launch into a riff on how calculator dependence has crippled America's youth, and when you see students multiplying 26 by 10 using a calculator instead of adding a zero, you know that society has problems. But we are where we are, and these problems will persist until we can make a compelling case for using our noggins instead.

We start by noting that for tens of thousands of business projections and feasibility studies undertaken each year, a rough, preliminary answer to the question at hand can be far preferable to a more exact answer. Exact answers cost time and money, and are often unnecessary to boot. If your company can't afford to spend $2 million on a new computer system, no one has to fine-tune the actual cost to $2,453,277.81. For the most part, we are in a game of fiscal horseshoes, where close counts just fine.

In certain circles, estimating and rounding skills remain highly valued. Job candidates at the prestigious Boston Consulting Group must face interview questions such as "How many pay phones are there on the island of Manhattan?"[25] The official BCG estimate was obtained by assuming that the island is 300 streets long and 10 avenues wide, for a total of 3,000 intersections. Assuming one pay phone for every two intersections, you get an estimate of 1,500. From there, BCG suggests making adjustments for the absence of intersections in Central Park, and for pay phones found indoors in restaurants, schools, hospitals, and office building lobbies.

The estimate of outdoor pay phones is, of course, extremely primitive. If the actual number of avenues turned out to be 15 instead of 10, that's a 50 percent gap right there. No doubt the estimate of 300 streets running the other way was on the high side, so you could get some built-in balance. Except then we'd have to reexamine the one-in-two-blocks assumption, because small numbers have a special potential to introduce error. If the ratio were one booth per block instead, you'd double the total in a heartbeat. At the other extreme, if you're nostalgic enough to believe that phone "booths" must be enclosed, your answer would be zero: AT&T stopped making enclosed phone booths in the mid-seventies, when vandalism costs hit $15 million a year.[26]

The point is that no one answer is correct. In real life, management consultants must evaluate market potential from the bottom up, and their world is filled with estimates and outright guesses. (For practice, try tackling a more challenging BCG question: "How many hotel-sized bottles of shampoo and conditioner are produced each year around the world?" And then move on to the question a friend received in a McKinsey interview: "How many disposable chopsticks are used in Japan daily?") In the same vein, securities analysts make earnings estimates based on models, and the growth assumptions that underlie their estimates can be far more important than the estimates themselves. This latter point is easy to overlook when Cisco shares are getting hammered because the company's quarterly earnings per share came in a penny below Street estimates. But the stock market is an esti-

mating system that only gets it right over time, which is why
growth stock investors learn to develop long-term orientations in
lieu of short-term ulcers.

Speaking of sensitive numbers, denominators are notoriously
fussy. For example, if you fill up your gas tank with 14.4 gallons
and your trip odometer registers 388 miles for the last fill-up, you
can round to 400 and 14 for an estimated mileage of 28.6 miles
per gallon. But in successfully applying grade school rounding
techniques (rounding 388 up and 14.4 down), you've moved the
numerator and denominator in opposite directions, which is asking for trouble. Dividing 400 by 15 gives 26.7, two miles per gallon less and a much better estimate of the actual figure of 26.9
miles per gallon. No great harm done here, but the smaller the
denominator is, the more error you'll create by rounding it off.

Curiosity is the hallmark of a quantitative thinker, and it goes
hand-in-hand with the empowering sensation that answers are always within reach. In turn, if you cannot question numbers, you'll
take what you get, even if Earth weighs only 3,000 kilograms as
a result. That's why good quantitative thinkers are always scrib-
bling in margins or on backs of envelopes. Inquiring minds want
to know, and they want to know fast.

Our ability to make quick calculations often depends on knowing which variables you need to take into account and how to get
at them. How fast do racehorses run? Is it twenty miles per hour?
Sixty miles per hour? If those questions occur to you as you watch
the Kentucky Derby, not only do you win points for curiosity, but
your curiosity should in turn give you access to all the information you need. Derby commentators invariably mention 1) that
the race is precisely a mile and a quarter long, and 2) that the
track record was set in 1973 by Secretariat, who finished with a
time of 1:59 $\frac{2}{5}$. Well, speed equals distance over time. Secretariat's time is just begging to be rounded off to two minutes,
which is 1/30th of an hour, so his speed must have been 30 × 1¼,

or 30 + $^{30}/_4$ = 37½ miles per hour. No fuss, no muss. No calculator used, and no calculator needed. The answer isn't exact, except in the sense that it gives you precisely what you wanted to know.

It's remarkable how often numbers fall into place if you'll just let them. Secretariat's Triple Crown–winning time for the Belmont Stakes was 2:24, exactly 20 percent more than the 2:00 we gave him for the Derby. And guess what? The Belmont, at 1½ miles, is also exactly 20 percent longer. So our earlier approximation of his speed is, in this case, exact. We already knew that Secretariat's performance in the Belmont was remarkable. He did, after all, win by 31 lengths. But now we know that he sustained his Kentucky Derby pace—a pace that no horse has ever beaten over the Derby's 1¼ miles—for another quarter mile.

We said that this section would make us appear old-fashioned, and maybe that's so. But we honestly believe that being able to make quick calculations is an important skill for any generation. You get answers, perspective, and confidence, all for relatively little work. So the next time you reach for your calculator, first try to come up with a ballpark estimate of your answer. And if the calculator comes close but insists on giving you seven decimal places, reprogram it.

Habit 10: Look for the Easy Way Out

For our final habit, you're going to get a break. Until this point, you have needed to keep your guard up at all times. We have asked you at various times to insist on numbers, to probe numbers, and even to distrust numbers. We have asked you not only to answer questions, but also to make up questions of your own. In recognition of all this hard work, we're now encouraging you to goof off.

Lest you be expecting a piña colada and a hammock, we should explain what we mean. Part of what makes someone a superior quantitative thinker is the ability to find the easiest possible approach to a number-driven problem. If you're always tackling problems the hard way, how can you stay alert and inquisitive when it matters the most? Better to save your energy. Besides, if you look for shortcuts, you just might find insights in the process.

Some years ago, one of your authors was watching a teacher explain how to calculate a grade point average. The assumption was that the student had gotten 3 A's, 1 B, and 1 C in his five courses. The calculation began pretty much as you'd expect. Each letter grade was given a number, as in 4 points for an A, 3 points for a B, and 2 points for a C. Adding up all the points yielded a total of $(3 \times 4) + (1 \times 3) + (1 \times 2) = 17$. At this moment, everyone in the class expected the teacher to divide 17 by 5, which in due course would have produced a GPA of 3.4. But that's not what happened. Instead he *multiplied* 17 by 2, getting 34 in one second flat, then stuck in a decimal point to reach the same result. The underlying principle was simplicity itself: If you want to divide a number by 5, you can instead multiply by 2 then divide by 10. You didn't have to be much of a mind reader to know that everyone in the class was thinking the same thing: "Why didn't *I* think of that?"

Shortcuts such as this one are seemingly less valuable today, now that calculators have taken over the business of calculating. But compare the effort to punch out all the above numbers on a calculator, versus doing it all in your head. Do we really feel that it's easier to undergo extensive physical labor than to use our brains? Hold that thought as you consider the classic tale of numerical energy saving, brought to us in the late 1700s by German prodigy Karl Friedrich Gauss. As the legend goes, Gauss was quite a handful as an elementary school student, so his teacher sought to keep him busy by challenging him to add up the numbers from 1 through 100, no doubt envisioning a long stretch of peace and quiet. We can sympathize with the teacher's lack of enthusiasm when Gauss returned just minutes later, correct answer in hand.

The beauty of Gauss's solution lay in his recognition that the numbers on his list paired themselves in a natural way: The pairing of 1 and 100 yielded 101, ditto for 2 and 99, 3 and 98, and so on. Altogether he had 50 pairs of numbers, each of which summed to 101, so it was no great stretch to conclude that the total sum was 50 × 101, or 5,050. Recognizing patterns beats the hell out of real work.

To someone who enjoys patterns, we might mention that 5,050 is the one-hundredth "triangular" number, where each such number is obtained by starting with one and adding successive whole numbers. Anyone who has ever gone bowling knows that 10 is a triangular number. So is 21, which just so happens to be the total number of pills in a "Med-Pak" of anti-inflammatories. The Med-Pak isn't a 3-week supply, it's a 6-*day* supply, in which the patient takes 6 pills the first day, 5 the second day, and so on. The point is that when you see patterns in the making, numbers become easier to explain and problems become easier to solve. If you don't believe us, ask the hapless souls who tried adding 1 through 100 on their calculators. They're still trying.

Even if you don't want to go head to head with Gauss, you can use a type of pairing technique anytime you have to add or subtract in your head. If you're down to $95 in your checking account and you write checks for $7 and $15, you could go 95 − 7 = 88 and 88 − 15 = 73, but any self-respecting quantitative thinker would be lazy enough to subtract the 15 first, getting 95 − 15 = 80 and then 80 − 7 = 73. While you'd be unlikely to go wrong using the first ordering, it's *almost impossible* to go wrong using the second ordering. And when you have $73 in your bank account, you can't afford to make mistakes.

Balancing your checkbook is of course a notoriously dreary task, so much so that many of us (or is it most?) take the ultimate easy way out—letting the bank statement and ATM do our work. Fine, but it is an annoying fact of life that as long as some of your checks have yet to clear, the ATM or statement balance won't tell you what you actually have in your account. If you keep detailed records and never make careless errors, you won't go wrong, but who in the world fits that description? The best approach for mere mortals is to be able to troubleshoot and solve problems as they come up.

Fortunately, some recurring problems have easy remedies. If your balance is off by precisely $100, the most likely culprit is an ATM withdrawal you never recorded. If that search comes up empty, chances are you'll be able to find a little carrying or borrowing problem in the hundreds column. On the other hand, if you're off by $2.50, $5.00, $7.50, or some other suspiciously familiar number like that, you've probably omitted one or more months of bank fees.

Other, more esoteric, problems can be resolved simply by checking out the last two digits of the discrepancy. If your ATM gives you a balance that appears, say, $51.39 too low, you might try looking for checks whose amounts end in 0.61, because it's possible that your manual balance was in fact $100 too high to begin with, and a check for the "complementary" amount of $48.61 hadn't cleared yet! And if you want to get truly esoteric, suppose that your calculations are off by the tantalizingly puny amount of 36 cents. It's entirely possible that your error was one of transcription—you could have written down $25.73 in your checkbook, for example, when the actual amount was $25.37. Why would you suspect such a thing? Because transcription errors are a prime suspect any time you are off by 0.09, 0.18, 0.27, 0.36, or basically any other multiple of nine cents; it so happens that the difference between a two-digit number and its reversal must always be a multiple of nine.

In general, calculations are rendered easier if you strive to keep all numbers as small as possible. If you're in Canada and you hear that the temperature is 20 degrees Celsius, you might want a Fahrenheit equivalent. The formula $F = \frac{9}{5}C + 32$ is your ticket, but don't start by multiplying 9 times 20 to get 180. Instead divide 20 by 5, multiply the result by 9, then add 32. Unless the cool Canadian air is getting to you, you'll get an answer of 68 degrees.

Even easier, you could use the stripped-down conversion formula we learned from TV's McKenzie brothers. To translate Celsius into Fahrenheit, the Molson-drinking Canadian pair simply doubles the Celsius figure and adds 30, whereby 20°C becomes 70°F. As an easy-way-out method, the McKenzie formula is a thing of beauty. It eliminates the fraction $\frac{9}{5}$, replaces 32 with a multiple of 10, and balances its overestimate of the $\frac{9}{5}$ with an underestimate of the 32, thus making the formula stunningly accurate. Between freezing 0°C and room-temperature 20°C, the McKenzie formula is never off by more than 2°F.

Estimating the arrival date of a newborn provides an unexpected opportunity for an easy solution. As this chapter was being written, the wife of one of your authors took a home pregnancy test and tested positive, and before spreading the word to the extended family, the parents-to-be wanted to provide a delivery date. Well, it's more guesswork than you might think. Books on the subject allow for the fact that 1) in the best of worlds, conception dates are ambiguous, and 2) sperm are slow, so even if you could pinpoint your conception effort to the day and the hour, you wouldn't necessarily know when fertilization occurred. That's why some books will tell you to estimate 40 weeks after the beginning of the mother-to-be's most recent menstrual cycle. But try flipping ahead 40 weeks on your wall calendar. It's more difficult than it sounds. Better to take the easy way out and *subtract* 12 weeks and a day. In doing so you've minimized your work and obtained a more reliable estimate, which may come in handy when relatives are breathing down your neck nine months later.

You should be aware that seeking a streamlined solution has the backing of at least seven centuries of philosophical thinking. The pet theory of British theologian William of Ockham (1285–1345) was that "plurality is not to be assumed without necessity," or "what can be done with fewer assumptions is done in vain with more." This precept, which endures under the odd name (with Latin spelling) of "Occam's razor," doesn't deny that the world's problems can be complex, but it recommends the pursuit of simple approaches and models whenever possible. When long, convoluted explanations have to be added to existing theories (as happened centuries ago when the geocentric theory of the universe began to break down), that's a clue that the original construct might have fatal flaws.

In an attempt to keep Occam's razor eternally sharp, we would add that easy solutions are most likely to crop up when you need a *negative* answer to whatever question you're asking. In that respect, positive inferences are to Dame Agatha Christie what negative inferences are to boy detective Encyclopedia Brown. The former identifies a murderer through a book-length investigation, using all available evidence, from motive and opportunity all the way down to mannerisms and appearances. However, in Encyclopedia Brown's grade school–oriented five-minute mysteries, the culprit must be nabbed within a few short pages. The inevitable result is that the entire case rests on a contradiction. Wait, that balloon was blown up by hand, not by helium. If the gun really was a six-shooter, how did it leave seven shots? And so on.

Effective calculations are the same way. If we've learned anything from habits such as *Go Figure* or *Live by Pareto's Law*, the affordability of a major household expense is never an issue of pennies. Yes, *after* you've qualified for a mortgage, it's true that the bank prefers your monthly check to be accurate to the cent. But if you're scoping out what sort of house you can afford in the first place, you can skip the decimal points, because if $1,500 a month is too much, then so is $1,533.16, and, for that matter, so is $1,487.83. Elsewhere along the same theme, if a vote passed the Senate by a 71–29 margin, the exact final tally might be of interest to the *Congressional Record*, but both sides probably stopped counting as soon as they could divine the outcome. Rounding and estimating should be in everyone's bag of tricks, for the simple reason that they make life easier.

To make sure you take the spirit of the easy solution with you as you leave this chapter, we will close with an urban legend of laziness done good. The setting is a physics exam, and the task is to determine the height of an apartment building using only a barometer. The preferred method is to apply the rule of thumb that near sea level, atmospheric pressure declines by about one millibar for every 10-meter increase in elevation, or about one inch of mercury for every 1,000 feet. But there is more than one way to skin this particular cat. As the story has it, a particularly clever student offered three alternative approaches.

Alternative method #1: Drop the barometer from the roof and count how many seconds it takes to hit the street below. Plug that time into the Newtonian formula $d = 16t^2$ to obtain the distance in feet.

The obvious flaw in this solution is that you still need to know some physics to pull it off. Of greater appeal would be the following:

Alternative method #2: Tie a long string to the barometer, go up to the roof, and lower the barometer until it hits the ground. Then measure the string.

Ah, but you still have to do some legwork, don't you? The truly lazy won't rest until they find the ultimate solution:

Alternative method #3: Find the building's superintendent and make him an offer. If he tells you the height of the building, you'll give him a brand-new barometer.

THE TEN HABITS OF
HIGHLY EFFECTIVE QUANTITATIVE THINKERS

Attitude Is Everything
1. Only Trust Numbers
2. Never Trust Numbers

Navigational Tools
3. Play *Jeopardy*
4. Live by Pareto's Law

Illuminating Numbers
5. Play 20 Questions
6. Build Models

Uncertainty
7. Play the Odds
8. Know What You Know and Don't Know

Estimation
9. Go Figure
10. Look for the Easy Way Out

For Good Measure

It's easy to forget that numbers don't appear out of thin air. They pop up everywhere, without any accompanying account of their origins. But numbers not only come from somewhere, they usually have a substantial history by the time we see them.

Think about the last number you saw before reading this book. Was it the time of day? A sum on a cash register? A page number in a magazine? Unless it was a phone number or a cable TV channel, chances are the last number you saw *measured* something. Most numbers are merely the end result of a multi-step measurement process: Somebody decides to measure something; that person does so according to certain standards, definitions, or specifications, chooses particular units, employs selected measuring techniques and equipment (none of them error-free); and then reports the result in some format, often after converting it into different units or rescaling it in other ways. Or at least that's the way it goes when things are simple, like when a newspaper tells you yesterday's high temperature.

The story behind a number is often as interesting and important as the number itself. That story is ignored at our peril, because the many steps involved in creating a number affect what information is captured in that number and how that number is likely to be interpreted and used. So if we really want to understand a number, we must take a hard look at how it was created.

Units 101

The first step toward understanding numbers is to recognize that numbers without units don't tell you much. (Numbers that order, such as page numbers, are exceptions.) If you stopped to ask directions of someone, only to hear that your destination was five away, surely you'd ask, "Five what?" Five blocks? Five miles? A five-minute walk? A five-minute drive?

Men may find this observation trivial—nobody would say "five" without giving units. Women know better. They see numbers without units every time they shop for clothes. Men's sizes are expressed in inches. A 42 jacket, for example, is sized to fit a man whose chest circumference is 42 inches. But the number 8 on a dress size doesn't quantify anything, which gives apparel makers wide latitude in scaling their sizes. Predictably, marketing considerations weigh heavily, and many clothiers believe it's profitable to flatter women by offering large garments labeled with small sizes, a practice known as "vanity sizing."[1] The result is striking inconsistency in sizing, which, if we can generalize from our significant others, drives women crazy.

Math books offer all sorts of guidelines for working with units, but, frankly, most of the precepts are so obvious that you have to wonder whether readers feel enlightened or insulted. Here's an actual sentence from a popular college-level quantitative reasoning textbook:

> You cannot add or subtract numbers unless they have the
> *same* units. For example, 5 apples + 3 apples = 8 apples,

but the expression 5 apples + 3 oranges cannot be simplified further."[2]

A point evidently not lost on Cézanne when he named his famous painting *Pommes et Oranges.*

In thinking about units, we propose starting with two simple
principles:

1. Quantities should be expressed in units that can be
   easily grasped.
2. Units should simplify calculations and comparisons.

Viewed individually, these principles seem every bit as obvious
as the one we just poked fun at. But putting them together can
produce all sorts of trouble. Just ask greengrocer Steve Thoburn,
who in 2001 was convicted by an English court of violating the
1985 Weights and Measures Act and given a six-month condi-
tional discharge. His offense? He sold a pound of bananas.

Thoburn asked for it. Weeks before his arrest, Thoburn, now
known to his countrymen as the "Metric Martyr," had been warned
that his imperial-unit scales infringed regulations by weighing in
pounds rather than kilos. A European Union directive, later
adopted into British law, requires that loose goods be sold in metric measures.

Forget that Thoburn should probably never have been charged
for such a harmless violation. What we find interesting about the
dispute is that Thoburn found himself trapped between the two
principles we just advanced. His customers clearly grasped pounds
better than kilos: Around that time, the British supermarket chain
Tesco commissioned a survey of its shoppers and found that 90
percent "thought" in pounds and ounces and only 8 percent preferred metric measures.[3] But the imperial system has always been

absurdly complicated relative to the metric system, a system whose conversions, whether between gram and kilogram or meter and centimeter, involve the simplest possible arithmetic: either adding zeros or taking them away. It's no wonder that the metric debate has flourished for as long as it has.

What's also interesting about the case is that if anyone is blameworthy, it's neither Thoburn, nor EU rule makers, nor members of Parliament. The bad guys are the British shoppers, because they neglected a basic quantitative responsibility. Just as those who create numbers (like Thoburn when he weighs bananas) have an obligation to use units that are readily understandable and easy to work with, consumers of those numbers have the responsibility to comprehend simple measurements expressed in units they can expect to encounter. Surely it's not unreasonable to ask UK citizens, who are also EU citizens, to be comfortable with metric measurements of grocery items. However, in the Tesco survey, 53 percent of shoppers admitted finding metrification confusing.

Here's how to avoid shirking the duty to understand common units. Remember the 20 Questions habit from the previous chapter? When a number shows up on your doorstep, the first question you should ask is, "What are the units?" This sounds like a no-brainer, yet it's easy to get careless. Don't. We'd say that someone could break his neck, but it's actually worse than that. On April 15, 1999, a Korean Air cargo plane departed from Shanghai's Pudong Airport. Shortly after takeoff, a Chinese air traffic controller directed the pilots to climb to 1,500 meters. The pilots, accustomed to altitude measurements in feet, interpreted the instruction as 1,500 feet and descended steeply. They recognized their error too late and the plane crashed, killing eight.[4]

The second question you should ask is, "How can I best conceptualize the units?" Although we won't cast this question as a matter of life or death, it's every bit as important as the first one. If you don't have a sense of what units represent, you could make the kind of mistake that two Harvard roommates made in 1981, during their first week of college. Excited about having a fireplace in their dorm room, they ordered a cord of firewood. They didn't know what a cord was, but they had been told it was the standard unit for firewood. True enough, but that standard unit turned out to be 4 feet high by 4 feet wide by 8 feet long. Picture that in a small college dorm room. On the plus side, residents of an entire Harvard dormitory had as much free firewood as they wanted.

Had the roommates simply consulted their handy *Merriam-Webster*, they might have avoided filling a third of their living room with musty old logs. In general, though, definitions are surprisingly lousy as a means of conceptualizing units. As an illustration, consider a garden-variety unit, the acre.

What Is an Acre?

"Acre" is a common enough term, so you'd think each of us would have a ready answer. But all we really know is that an acre is a measure of land. Our instincts tell us that a 40-acre plot is sizable and that a half-acre plot is not. Beyond that, the acre remains surprisingly ill-defined.

Enlisting a dictionary provides us with numbers, but not perspective. We learn that an acre is 43,560 square feet, which happens to be equivalent to 160 square rods. There's even a metric tie-in: An acre is very close to four-tenths of a hectare, which in turn is 10,000 square meters. Sigh. If our goal is to develop a feel for what an acre really is, the dictionary compilers haven't helped us much. Perhaps they were saving their strength so early in the alphabet.

The reason "acre" is so slippery is that it is a generic creature, much like measurements of sound and volume such as "octave" and "decibel." When we say that fright caused someone's voice to go up an octave, we aren't suggesting that the difference in pitch was precisely eight diatonic degrees. We simply lack alternative descriptions within the musical scale, so "octave" gets a little worn out. "Acre" is the same way. In isolation, it seems arbitrary. To the settlers of the Old West, the concept had specific meaning; since biblical times, an acre had been recognized as the section of land that a yoke of oxen could plow in one working day. There were reckoned to be 640 such sections in a square mile. But our modern-day intuition is still somewhat wanting.

When you think about it, what makes the acre stand out as a measure of area is that we have no sense of its *linear* dimension— in contrast to terms such as square inch, square foot, or square mile. And just because we know what a square foot is doesn't help our appreciation for what 43,560 of them might amount to. So a natural approach is to convert an acre into its corresponding linear measure. In other words, if an acre were marked out as a square, how big would that square be?

For the wise guys who shouted out, "The square root of 43,560," you're right, but we'll settle for an approximation. Note that 43,560 is just a touch more than 40,000, a number that consists of a perfect square followed by an even number of zeros. Taking square roots of such numbers is child's play: Simply take the square root of 4 (2, on a good day) and add half the number of zeros in 40,000 (again, 2). What we're left with is the estimate that an acre of land is a trifle larger than a square plot measuring 200 feet by 200 feet. Finally we have something to which we can relate.

Two hundred feet is of course two-thirds of the distance from goal line to goal line on a football field, so imagining that distance squared gives us an image worth retaining. Alternatively, consider that the entire playing area of a football field (not including the end zones) is 100 yards by 160 feet, or 48,000 square feet—roughly 10 percent more than an acre. A caveat here is that our sense of a football field's size may be swelled, in that our eye takes in the entire out-of-bounds area as well as the playing area itself.

To make sure we don't leave any loose ends, note that the number of square feet in a square mile equals 5,280 times 5,280, or 27,878,400. Taking 1/640th of that, you get 43,560, a number that apparently wasn't picked out of thin air after all. Perhaps the oxen of the Old West should have been speedier, but we can take comfort that the settlers' arithmetic was right on target.

Conversions: Putting Units Together

If we understand an acre better than before, fine. But we aren't going to push our luck by dissecting such units as watts, amperes, joules, and ergs. Instead we will stick with familiar units, this time focusing not on any one unit, but on the all-too-common challenge of maneuvering between different ones.

We know that we can pay a price for not making the proper conversions. NASA scientists discovered as much in October 1999, when they lost their $125 million Mars Climate Orbiter to a metric snafu. Thrust data sent by Lockheed Martin were delivered in the British force unit of pounds, whereas NASA's computers were expecting numbers calibrated in the metric unit, newtons. As a result, the spacecraft, after traveling some 400 million miles, went about 50 miles too far, and came within 36 miles of the planet's surface—or 58 kilometers, as reported by the news media in an apparent NASA jab. Whatever the units were, the Orbiter overheated and was lost in the Martian atmosphere.

On an individual level, we suffer when numbers come to us

without recognizable benchmarks. When Bob Beamon leaped 29'
2½" in the 1968 Olympic Games—thus beating the prior world
record in the long jump by almost two feet—he knew he had
done something special the instant he landed, but he didn't immediately know just *how* special it was. The electric scoreboard
took its sweet time and eventually popped out "8.90," as in meters, a number that had no resonance whatsoever. Only when
teammate Ralph Boston clued him in—"Bob, you jumped 29
feet!"—did the moment set in.

The point is that when numbers aren't delivered to us in the
form we'd like, we had better be able to convert them into units
we can readily comprehend. That's why textbooks and other reference works provide tables that look something like the following smorgasbord of conversions.

1 mile = 5,280 feet	1 ton = 2,000 pounds	1 gallon = 4 quarts
1 yard = 3 feet	1 pound = 16 ounces	1 quart = 2 pints
1 foot = 12 inches	1 stone = 14 pounds	1 tablespoon = 3 teaspoons
1 mile = 1.609 kilometers	1 pound = 0.454 kilograms	1 gallon = 3.785 liters
1 kilometer = 0.621 miles	1 kilogram = 2.205 pounds	1 liter = 0.264 gallons

The list is mercifully incomplete. We haven't mentioned that a
liquid quart is different from a dry quart. Or that a pint of beer in
London—half an Imperial quart—is 20 fluid ounces, not 16, and
that each fluid ounce in London is 3.92 percent smaller than its New
York counterpart, the net effect being that a pint of beer in London
has 20.1 percent more beer than a pint of beer in New York.

Having conversion numbers at our fingertips is a beginning,
not an end. If some of the above numbers are familiar, great, but
we aren't suggesting that entire conversion tables be memorized.
First of all, unless you're trading foreign currencies in large de-
nominations, you almost never need conversion factors carried out
beyond one or two decimal places. Second, any conversion can be

derived if you know the reverse conversion: for example, if you remember that a kilogram is about 2.2 pounds, then you can easily figure out how many pounds are in kilogram. To do so, you simply divide 1 by 2.2, obtaining 0.45. When you divide 1 by a number, you get what is called the reciprocal of that number. Within our conversion table, each number on the bottom row is the reciprocal of the number directly above it.

Reciprocals are simple in concept, but their hidden role in unit conversions can create massive confusion. Currency conversions are a prime culprit, for world travelers and the print media alike. For example, the July 10, 2000, issue of *Business Week* contained an article titled "The Peso Is Falling! The Peso Is Falling!"[5] And so it was, but you wouldn't have known it from the graphic. The line in the chart at the bottom of the page was going *up*, not down. The discordance came about because the data were expressed in Mexican pesos per U.S. dollar, a figure that *rises* as the peso becomes less valuable. Making the chart point in the proper direction would have required taking the reciprocal of the data points, obtaining U.S. dollars per peso. The blunder was reminiscent of what a finance minister said of the struggling Japanese economy in early 2001. "It looks okay, as long as you turn the charts upside-down."

Speaking of hidden conversions, here's a disarmingly simple problem. It began when Cliff Sahlin, a teacher at the Foote School in New Haven, Connecticut, ran in a four-mile race, finishing with a time of 26:48. Not exactly Roger Bannister, but anything under seven minutes per mile is quite respectable for the masters division. When Sahlin returned to work, he asked his class to calculate his precise time per mile. Many of the students came up with a figure of 7 minutes, 2 seconds. Impossible, right? If he had averaged over seven minutes per mile, then his total time would have exceeded 28 minutes. So where did the mistake come from?

When you do the arithmetic, you see that you need to account for the fact that clocks don't run on the decimal system. Students who didn't bother with converting ended up dividing 26.48 by four, obtaining 6.62, which they then read as 6 minutes, 62 seconds! An honest mistake, but you can be certain that Sahlin pointed out the red flag. The simplest approach was to divide 26 by four, obtaining $6\frac{1}{2}$, then divide 48 seconds by four, for an extra 12 seconds. Adding those 12 to the $6\frac{1}{2}$ minutes you already had, you get a total of 6 minutes, 42 seconds, a better answer than six point six two minutes, any day.

Sometimes conversions enable us to visualize a situation that otherwise would be too abstract. Recall from chapter 2 that Secretariat won the 1973 Belmont Stakes in 2:24, the fastest Belmont time since the race was lengthened to a mile and a half in 1926. Less well known is the fact that in 1989, Easy Goer recorded the second-best Belmont time ever, at 2:26, a time then equaled by A. P. Indy a few years later. That's a two-second differential, but the comparison would be much more vivid if we knew the *distance* represented by a two-second gap. That's where conversions come in.

Our first step is to take our prior calculation of Secretariat's average speed—$37\frac{1}{2}$ miles per hour—and convert it into feet per second. We start by noting that there are 5,280 feet in a mile, so $37\frac{1}{2}$ miles per hour equals $37\frac{1}{2} \times 5,280$ feet per hour. But there are 3,600 seconds in an hour, so to obtain the number of feet traveled each second we must divide by 3,600.

Remember, though, we'll take the easy way whenever possible, and we have no intention of carrying out the full calculation $37\frac{1}{2}$ $\times \frac{5,280}{3,600}$. Instead we observe that the $37\frac{1}{2}$ figure is just an approximation. It could be smaller, because the relevant speed is that of Easy Goer and A. P. Indy, who were traveling a tad slower than Secretariat. Or it could be larger, on the assumption that a horse in a stretch run is running faster than his average rate. Whatever the exact number is, it is just a smidgen more than the 36 in the denominator. The number we're interested in is therefore very close to $36 \times \frac{5,280}{3,600}$, and once we cancel out the 36's and drop a couple of zeros, we're left with 52.8. The units belonging to that number are feet per second, so if Easy Goer and A. P. Indy were two seconds off Secretariat's pace, they were more than 100 feet behind. That's the image we were looking for.

Unfortunately, there are occasions when all the conversion skills in the world won't get you the number you're interested in. In February 2001 an Associated Press story reported that ski races at the World Alpine Championships in St. Anton, Austria, had been postponed because "a 36-hour snowstorm dumped about 100,000 tons of snow on the slopes." How deep was the snow? Who knows? Without figures for the acreage involved and the density of the snow—was it light and fluffy or heavy and wet?—it's not possible to convert a weight of snow into a depth.

As an aside, the Associated Press isn't alone in having difficulty reporting snowfall figures. According to *Guinness World Records 2001*, "Between Feb. 1971 and Feb. 1972 a record 12,246 in. of snow fell at Paradise, Mt. Rainier, Washington, USA, a record for a single year."[6] If you think about it, that number can't be correct. Paradise may be the snowiest place on Earth, but there's no way it ever gets a thousand feet of snow in a year. We'll leave it to you to throw stones at the *Guinness* staff, who must have inadvertently added a digit to the actual record. We'll retreat to the dubious safety of our glass house, knowing that this book is almost certainly not mistake-free, despite our best efforts to make it so.

If you're into backpacking, you might be familiar with our final example of conversions. Conversions are necessary here because the activity of backpacking is more universal than the capacity measurements of the packs themselves. The volume of American backpacks is expressed in cubic inches, whereas Europeans use liters. So we begin with the conversion ratio of 1 liter = 61.024 cubic inches.

A clumsy number such as 61.024 underscores the point that the conversion process is often more decimal-happy than it needs to be. One regrettable outgrowth is that many backpacking books and articles take pride in their precision and therefore use the full 61.024 figure, or even its frightful reciprocal, 0.01639. At risk of merely confirming the obvious, no one really needs the volume of a backpack measured with such precision. *Backpacker* magazine recognizes that estimation will do very nicely, writing in its annual gear guide: "To calculate cubic inches, multiply the liter rating by 60. For example, a 70-liter pack has roughly 4,200 cubic inches of capacity ($70 \times 60 = 4,200$)."[7] Give them credit for a sensible use of approximation, although it's sad that the sentence didn't end after the word "capacity."

Taking the Measure of Measurement

If measurement isn't the most neglected topic in quantitative education, it's a strong contender for that designation. We looked at several basic math and quantitative reasoning textbooks aimed at humanities majors, and in none did the index include an entry for "measurement." This omission would be forgivable were it not for four inconvenient facts, the last one of which turns out to be a whopper:

1. There is always more than one way to measure something.
2. Measurements are error-prone.
3. Even when dead-on, measurements are often just an approximation for what you really want to know.
4. Measurements change behavior.

Our basic gripe is that most math books take numbers as given, and so do most of us. But these four observations add to the underpinning of this entire chapter, which is that you shouldn't read much into a number until you consider the measurement process that generated it.

In exploring this issue, let's pick up where we just left off, with some advice from *Backpacker* magazine about measuring the size of backpacks. "Until pack manufacturers come up with a universally accepted method for measuring the capacity of their products—they currently use everything from sand to dry dog chow, with pinto beans and Ping-Pong balls in between—take capacity figures with a grain of salt."[8] Measuring a backpack's capacity with pinto beans produces a higher volume figure than measuring with Ping-Pong balls. Pinto beans can fit into nooks and crannies that Ping-Pong balls cannot. And because pinto beans are denser and can be stuffed into a pack without crushing, they stretch pack fabrics, further increasing capacity figures. More important, though, backpacks have a variety of designs (in particular, some external frame packs are designed to carry big items outside the pack bag), and many items that get crammed into backpacks—among them tents, stoves, sleeping bags, and hiking clothes—aren't exactly shaped like pinto beans or Ping-Pong balls. What you really want to know is "How much of my stuff will the backpack hold?" Which leads *Backpacker* to propose a simple non-quantitative solution to the problem of backpack measurements: "Bring all your gear to the store to ensure that everything can be squeezed inside the pack you're planning to buy."[9]

No doubt, lugging all of your gear to outdoor equipment stores and loading it into backpacks gives you much better information about pack capacities than do manufacturers' volume measurements. But it's a hassle. And it's also conspicuous enough to attract onlookers. The larger lesson here is that better information often comes at a cost. Sometimes that cost is worth it, sometimes it isn't. Either way, the key as a quantitative thinker is to be aware of what information is and isn't captured by whatever measurement method is used.

Do you need to lose weight? Beauty may be in the eye of the beholder, but when it comes to body weight, physicians have a standard quantitative method for determining if you have too much of it. Your body mass index (BMI) equals your weight in kilos divided by the square of your height in meters. (If you're in metric denial, use pounds and inches and multiply the result by 703—or by 700 if you're doing the arithmetic in your head.) According to National Heart, Lung, and Blood Institute guidelines, a BMI of 25 is considered overweight, and a BMI of 30 is obese.

Now wait a minute. High body mass index is not in itself a major health risk, at least not compared to excess fat. To be sure, those with higher BMIs *tend* to be fatter than those with lower BMIs, but the correlation is far from perfect. Because some people have larger frames, heavier bones, or bigger muscles than others of the same height, BMI is influenced by factors other than fatness. Doctors have known as much since at least the early 1940s, when Navy physician Albert Behnke pioneered the measurement of body fat. Behnke and his colleagues evaluated 25 football players, 17 of whom had been declared unfit for military service on account of excess weight. The researchers determined that many of the athletes were lean and that their "excess" weight was attributable to big muscles.[10] In other cases, BMI guidelines can overlook fatness. Small-framed individuals with little muscle mass can have too much fat and still sport a BMI below 25.

So if BMI is an imperfect gauge of fatness, why use it? Why not just measure fatness directly? Simple. Height and weight can be accurately measured in less than a minute using inexpensive equipment that just about everyone has at home. Not so for body fat percentage, the customary yardstick for fatness. Just look at the names of some of the techniques used to assess body composition—hydrodensitometry, air displacement plethysmography, isotope dilution, dual-energy X-ray absorptiometry—and you'll rightly suspect that things are more complicated than stepping on a scale and standing next to a ruler.[11] Take hydrostatic weighing, long considered the gold standard for estimating body fat percentage. In a procedure reminiscent of a carnival dunking game, subjects are typically weighed 8 to 12 times while submerged, each time holding their breath for 5 to 10 seconds after exhaling completely. The resulting density measurements are then used to estimate body fat percentage. All that work and no stuffed animals.

We're leading up to a huge problem here. Since weight is easy to measure and fat isn't, our attention inevitably gets focused on weight and not fat. Reinforcing this misplaced focus is society's preference for euphemistic language. It's less insulting to say "You should lose some weight" than "You're fat." An estimated 50 million Americans will go on a diet this year, and most will gauge their progress in slimming down by weighing themselves. For those who don't understand that weight loss is a flawed measure of fat loss, getting on a scale every morning can provide counterproductive information, at least in the short run.

Many well-designed health programs combine moderate caloric restriction with regular exercise. In the near term, such programs may result in little if any weight loss, as the muscle mass gained from exercise offsets lost fat. Diet and fitness expert Covert Bailey reports that previously sedentary women sometimes quit such programs despite dropping a clothing size or two simply because they are discouraged by a lack of weight loss. These women have traded fat for muscle, a winning exchange that unfortunately doesn't register on a bathroom scale.

Bailey observed an illustration of a much larger phenomenon.

"What gets measured gets done," or so goes the old saw. One reason that measurement is so important is that measurements are used to assess and influence performance. We give exams to students to see what they've learned, but also to push them to study harder. However, whether we're talking about multiple-choice questions versus essay formats, or bathroom scales versus hydrodensitometry tanks, there's no perfect way to measure performance, so measurement inevitably influences behavior in both desirable and undesirable ways. If hospitals are required to publicly report complication rates for various procedures, doctors and nurses will have an incentive to take more care with patients. Terrific, but the same outlook might encourage them to avoid treating sicker and older patients who are at higher risk for complications.

Because measurement influences behavior, anyone who decides to measure something should think hard about unintended consequences. In an effort to discourage flight delays, the Department of Transportation requires airlines to submit data on departure and arrival times. Under the relevant regulations, a flight is deemed "on-time" if it "arrives less than 15 minutes after its published arrival time."[12] Never mind that most travelers don't consider 14 minutes and 59 seconds late "on-time." A bigger issue is that the regulations enable airlines to improve their on-time performance simply by lengthening scheduled flight times. Which is precisely what they've done. In Senate testimony, the Department of Transportation's inspector general stated:

> Between 1988 and 1999, the 10 major air carriers reporting to [the Bureau of Transportation Statistics] increased their scheduled flight times on over 80 percent of their domestic routes (1,660 of 2,036 routes). By increasing the schedule time, the actual extent of delays

through the system is underreported. For example, the number of arrival delays would have increased by nearly 25 percent in 1999 if the air carriers' scheduled flight times had remained at their 1988 levels. *We estimate that, from 1988 through 1999, these schedule changes added nearly 130 million minutes of travel time for air passengers.*[13]

Consider another federal policy involving transportation. The Corporate Average Fuel Economy (CAFE) standards promote energy conservation by requiring that the fuel economy ratings for all passenger cars sold by a manufacturer average at least 27.5 miles per gallon. (There's another CAFE standard for light trucks.) What's the adverse incentive here? Well, the CAFE standards encourage production of smaller and lighter vehicles, which are less safe than larger, heavier ones. But there's also an adverse incentive created by the measurement method. As a simple example illustrates, the CAFE standards, by targeting mileage ratings, do less than they might to reduce fuel consumption.

Suppose that among a carmaker's fleet are two vehicles, Model A and Model B, sold in equal numbers. Model A gets 20 mpg; Model B gets 40 mpg. From the perspective of the manufacturer, who is trying to satisfy the CAFE standards, increasing the fuel economy of Model B to 50 mpg is twice as valuable as raising Model A's to 25 mpg. But from an energy conservation perspective, it would be twice as valuable to make Model A more fuel efficient. Csaba Csere, editor-in-chief of *Car and Driver* magazine, explains:

> The argument is that if you drive 10,000 miles per year in a vehicle that gets 20 mpg, you burn 500 gallons of fuel. Raise the mpg to 25, and you use only 400 gallons, for a net saving of 100 gallons. Raise a car from 40 to 50

mpg, and you drop your fuel consumption from 250 to 200 gallons, saving only *half* as much fuel.[14]

By all means, try to be as smart as possible when measuring things. But most of the time, you will have to work with other people's measurements, rather than create your own. So you need to remember that every measurement method has strengths and weaknesses. If you can train yourself to identify those strengths and weaknesses, you will become a much better quantitative thinker. In fact, although you might not become a better butcher or candlestick maker, you have a chance to become a better baker, as we will now explain.

If you looked at a basic bread recipe in an American cookbook, you would find an ingredient list like this:

$3\frac{1}{2}$ cups flour
$1\frac{1}{3}$ cups water
2 teaspoons salt
1 teaspoon yeast

The flaw with this recipe is that flour is compressible. There's no standard method for measuring flour and many cookbooks do not specify the measurement technique appropriate for its recipes. (Given space considerations, newspapers and magazines almost never do.) In his book, *The Dessert Bible*, Chris Kimball identifies eight distinct methods of measuring flour:[15]

Sift and then spoon into measuring cup
Sift and then dip and sweep
Sift directly into measuring cup
Sift and then pour into cup
Spoon from bag; unsifted

Dip and sweep from bag; unsifted
Spoon from a canister; unsifted
Dip and sweep from a canister; unsifted

You should know that Kimball is an empiricist by nature, and he delights in tinkering with his recipes until he comes across just the right combinations. In this case he discovered that among the different measuring methods, the resulting weight per cup ranged from 3.1 ounces (sift directly into measuring cup) to 4.3 ounces (dip and sweep from bag; unsifted). Even among the unsifted approaches, weight per cup varied from 3.65 ounces to 4.3 ounces, enough of a discrepancy to significantly affect the bread.

We're confident that trial and error could resolve the problem, but contrast the bread recipe with how a comparable recipe might appear in a European cookbook:

500 g flour
320 g water
10 g salt
5 g yeast

The drawback with this recipe is that not everyone owns a scale precise enough for such measurements. If you do, however, measuring by weight is clearly an improvement. (Unless you store your flour next to your shower stall, where it will absorb enough moisture to approximate plaster of Paris.) Measuring by weight is the approach taken by many packaged goods companies, and you probably recall times when you opened what appeared to be a less-than-full box of cereal, cookies, or crackers, only to read the small print and discover the famous disclaimer, "Package sold by weight, not volume." In the absence of a good scale, the next best thing is a set of cups and spoons with metric measurements. But there's really no need for books to force us into buying extra apparatus. Some baking books provide recipes in multiple units. Rose Levy Beranbaum's *The Pie and Pastry Bible* and Charles Van Over's *The Best Bread Ever*, to name two, present each recipe in three versions: U.S. volume, U.S. weight, and metric weight.

But we're not done with our bread baking just yet. Our last step is to look at a professional bread-baking book, where we find yet another set of units:

Flour	100%
Water	64%
Salt	2%
Yeast	1%

The recipe is given in what is called "Baker's Percents," where all quantities are expressed by weight as a percentage of the flour. This approach has two significant advantages. First, the recipe is easily scalable—however much dough you make, you'll have no trouble keeping the proportions right. Second, attention is drawn to moisture content, one of the most important characteristics of bread dough. To a serious bread baker, that 64 percent figure for water (commonly termed "hydration") stands out like a batting average to a baseball fan—it tells the baker a lot about the resulting bread. A dry dough, say for a bagel, might have a hydration level of 50 percent, while the dough for a rustic-style loaf (the kind with big holes in the crumb) might be as wet as 80 percent. To a skilled bread baker, a difference of as little as two percentage points can be significant.

One Size Fits All?

We now move to a different type of conversion: new numbers formed by synthesizing old ones. Many numbers that we take for granted—among them batting averages, the Consumer Price Index, and even Microsoft's earnings—are built by simplifying and distilling a wide assortment of data into a single number. The hydration figure of the previous section was created in much the same way, and it was all to the good. Why juggle a dozen numbers in your head when a simple proxy is available?

When you think about it, the "single number" approach appears to have society's blessing, in that we routinely create fame by trying to identify the star within a crowd. The Supremes became "Diana Ross and the Supremes." *The Tonight Show* became *The Tonight Show starring Johnny Carson*. The problem in real life is that we often pay a price for fame. That risk is magnified in the world of numbers, because the advantages of simplicity can be more than offset if we don't pay attention to how these distilled numbers came to be.

For example, the wind chill factor was developed to convey in a single number how cold different combinations of temperature and wind really feel. Pick up an old wind chill chart lying around your house and you'll see that a temperature of 15°F combined with a wind speed of 15 miles per hour produces a wind chill equivalent temperature of –11°F. Thus, those conditions would feel the same as –11°F with still air. Right? Maybe if you were walking around buck-naked. The wind chill formula that generated your chart (which converts the two variables of temperature and wind speed into a single wind chill equivalent) approximates the convective heat loss of naked skin in the wind. But once you wear anything more wind resistant than, say, fishnet stockings, convective heat loss diminishes dramatically. Which means that for anyone other than a stripper booked for a December tailgate party, your chart would greatly overstate the effect of the wind.

The root of the problem is that your chart was based on the so-called Siple formula, created from the research of Brian Siple and Charles Passel, who in the early 1940s exposed plastic cylinders of water to the Antarctic elements and observed how quickly the water froze in different conditions. The Siple formula, critics argue, does not properly account for the myriad ways in which the human body differs from a plastic container. Nor does the Siple formula adjust for the fact that wind speeds reported by the National Weather Service, measured by exposed weather stations at an elevation of 10 meters, usually exceed wind speeds at ground level in ordinary human environments. With these criticisms in mind, the National Weather Service adopted a new formula, and as of November 1, 2001, conditions of 15°F with 15 mph of wind produce a wind chill equivalent of 0°F instead of −11°F. That the wind chill factor could rise by 11 degrees in the minute after 11:59 P.M., October 31, 2001, with no change in the actual weather, reinforces the point that such catchall numbers are inevitably imperfect.

In baseball, a batting average is the most common measure of a hitter's effectiveness. As a statistic, a batting average is as simple a quantity as you could imagine, obtained by dividing number of hits by number of at-bats. But even such a simple construction involves some hidden understandings. Walks, for example, don't count as at-bats; neither does getting hit by a pitch. On the other hand, getting on base courtesy of an error—or a fielder's choice, as when your ground ball retires your teammate at second instead of you at first—*does* count as an at-bat.

Once you recognize these factors, you also recognize that a batting average cannot tell you everything you need to know. If you were the agent of, say, Rickey Henderson, you'd have noticed that your client was especially good at drawing walks (he holds the all-time record), so you wouldn't base your negotiations on his batting average alone; fortunately, a figure called "on-base percentage" came along to solve your problems. Similarly, if you were representing Sammy Sosa, you'd know that his batting average treats his prodigious home runs as if they were mere singles. A slugging percentage, which divides total bases by at-bats, thus giving home runs four times the weight of singles, is more suited to your cause. Then again, slugging percentage, like batting average, overlooks walks, so many baseball observers look at "on-base plus slugging" or "on-base times slugging." But don't all of these measures ignore the value of stolen bases, or the cost of outs resulting from getting thrown out, picked off, or grounding into a double play? Yes, which prompted sportswriter Tom Boswell to champion "total average," the ratio of all the bases a player accumulates to the outs he generates. The search goes on and on, because no single number is perfect.

Yet all of these baseball measures are as simple as 1-2-3 when compared to some of the numbers we encounter outside the sports world. Consider the unit "net income," as for a corporation. Net income is defined by a labyrinth of rules, rules whose arcane details can dramatically affect outcomes. For example, when Daimler-Benz (now Daimler Chrysler) adopted U.S. accounting rules in order to get listed on the New York Stock Exchange, its "net income" dropped by more than $1 billion.

The underlying problem is that defining "profit" in a way that consistently and appropriately captures the financial performance of thousands of dissimilar firms is really, really hard. Saying, as an Economics 101 textbook would, "profits equal total revenues minus total costs" just doesn't get you very far because costs aren't the same as cash outlays and are thus difficult to define. Take the cost of equipment investments. When a widget manufacturer buys a widget-making machine for $100, it hasn't added $100 to its costs. It has rearranged its balance sheet, replacing $100 in cash with a machine worth $100. The cost of the widget machine is borne over time as the value of the widget machine declines. Let's say that after one year the widget machine is worth only $90; then the company has borne a depreciation cost of $10 in that year.

In theory, accounting depreciation should equal actual deprecia-tion. But having companies assess the fair market value of every-thing they own every year isn't feasible. It would entail an enormous amount of work and present an open invitation to cheating. So ac-counting rules provide different depreciation schedules for different kinds of assets. Inevitably, this several-sizes-fit-all approach has lim-itations. For years, cable television firms reported "losses," even though they were doing quite nicely, because they had incurred great "costs" from depreciating their infrastructure investments (ca-bles, etc.). In truth, their infrastructure was getting more valuable, not less, or at least it wasn't declining as rapidly as accounting rules allowed the companies to claim.

There's nothing revolutionary in these lines of thinking, but if something as simple sounding as "net income" or even a batting average can contain a few wrinkles, you can imagine the caveats associated with a number such as our very next stop: the Con-sumer Price Index.

Would you want the responsibility of devising an index to mea-sure inflation? It's a daunting task. Among the myriad difficulties is that people's shopping habits change in response to prices. Let's say that orange juice is one of the items in the "basket" of goods the Bureau of Labor Statistics uses to track prices. Suppose the price of orange juice then jumps. Many consumers will switch to close sub-stitutes—say grapefruit or pineapple—that have become cheaper in comparative terms. Therefore, the change in price of the basket with orange juice overstates the impact of inflation on consumers, since it gives too much weight to a product (OJ) that consumers now use less of.

Worse, there's the challenge of adjusting for quality. In 2001 the MSRP for a new top-of-the-line Honda Accord EX was $25,100. Back in 1991 the sticker price of an Accord EX was $19,050. It would be misleading, though, to attribute all of the $6,050 differ-ence—or possibly any of the difference—to inflation. Much or all of the increase in price can be chalked up to improved quality. "They don't make 'em like they used to," we often hear. Thank goodness. Violin quality may have peaked with Antonio Stradivari, but most products have gotten better, not worse. When was the last time you called a TV repairman? The trouble for the CPI is that it is difficult to measure the magnitude of quality changes.

The end result is that the CPI overstates inflation. According to the report of the 1996 Boskin Commission, that overstatement amounted to 1.1 percent per year. If this seems like a lousy punch line, consider that roughly 30 percent of federal government outlays are tied to the CPI, including federal and military pensions, indexed veterans benefits, and, most important of all, Social Security. The poverty line is linked to the CPI, as are income tax brackets and certain forms of tax credits and deductions. At the time of the Boskin Commission, it was estimated that shaving a percentage point off the CPI growth rate would reduce the 2006 federal budget deficit by a third, or $134.9 billion. The Bureau of Labor Statistics has since updated the CPI methodology to reduce some of the biases noted by the Boskin Commission, and one of the commission's members estimates that the bias has been reduced to around 0.65 percent per year.[16]

The point is that the CPI is a number whose creation requires constant vigilance. Like so many homogenized "single" numbers, it has more pressure on it than it can bear. Perhaps that's the true price of fame.

Bias and Precision

"If you have never baked by weight, borrow a scale and try it just once. I guarantee you will be an instant convert."[17] So claims professional baker Rose Levy Beranbaum. "There is no doubt about it: weighing is faster, easier, and more accurate than measuring."[18]

Since we're teaching you to be hypercritical in quantitative matters, let's look more closely at that last sentence. Faster, easier—those terms are clear. But what, exactly, does "more accurate" mean? This apparently simple question may be the hardest one we have asked so far, so bear with us if our answer is tough going. It's important.

When we measure something, our goal is to measure the true or correct value. This is an elusive goal, and normally our measurements are "off" by some amount. The technical term for "off" in this case is "measurement error," which is the difference between a measured value and the true value. If a surveyor estimates that a plot of land is 105 acres, and in fact it's exactly 100 acres, there's a measurement error of 5 acres.

Now for the hard part. There are two distinct ways in which a measurement can be off. A measurement can be biased (creating what is called systematic error) or it can be imprecise (causing random error). Suppose that the incompetent surveyor measured many 100-acre plots. If his acreage estimates were regularly off, but evenly distributed about 100, with just as many 98s as 102s, we would say that his measurements were unbiased but lacked precision. Conversely, if the surveyor consistently measured every plot to be 105 acres, we would say that his measurements were biased but precise. To put this another way, bias involves predictable measurement error; imprecision involves unpredictable error.

If you don't want to remember all this jargon, that's okay. Jargon confuses as often as it clarifies. What you do need to remember is the basic point that there are two different components of measurement error. When you encounter a measurement, you should wonder by how much it is off, and how much of any measurement error is attributable to bias as opposed to imprecision.

Back to backpacks for a moment. A couple of years ago, *Back-*

packer magazine compared six different medium-sized packs. All were heavier than their manufacturers claimed, suggesting systematic error rather than random error. At the extreme, the Rokk Neve weighed 5 lbs 6 oz on *Backpacker*'s scales, 48 percent more than the 3 lbs 10 oz manufacturer specs.[19] One of us subscribes to *Backpacker* and can attest that rarely is tested equipment lighter than the manufacturer claims. The folks at Rokk, however, seem to have particular trouble with their scales. Or do they? More recently, the magazine evaluated the Flat Iron, a pack that weighs 3 lbs 3 oz according to Rokk. *Backpacker*'s measurement, 4 lbs 12 oz, was 49 percent heavier.[20] It would appear, then, that Rokk's scales are very precise but highly biased.

Obviously, bias is not much of a problem if you know what it is. Any reader of this book in the market for a backpack now knows that the weight of a Rokk pack can be determined by adding half to the manufacturer's estimate. Bias is also not much of a problem when you are trying to measure changes in something, rather than absolute values. If you want to know how much weight you've lost after dieting for three months, it doesn't matter if your scale consistently flatters you by 5 pounds, because the difference between measurements isn't affected. But if your scale is imprecise, sometimes off by a few pounds on the high side and sometimes erring on the low side, then it's time to get a new scale. Better yet, spend your money on a pair of skin-fold calipers, so you can measure your body fat.

Beware of Strangers Bearing Zeros

On June 14, 1991, Leroy Burrell ran 100 meters in 9.90 seconds, establishing what was then a world record. But why 9.90? Why not spare some ink and write 9.9 instead? Because 9.90 tells us that Burrell's time was recorded in hundredths, and not tenths, of a second. A time of 9.90 does not equal 9.92 (the previous world record) or 9.86 (the subsequent world record), while 9.9 could equal either. The digits used to express a number's precision are collectively called "significant digits" (or "significant figures"). The number 9.90 has three significant digits; 9.9 has two.

If you look at the nutritional label on a box of Cheerios in a Detroit supermarket, you'll see that a 30 g serving has 3 g of fiber and 1 g of sugars. But pop across the border to Windsor, Ontario, and you'll find that a 30 g serving of Cheerios has 2.7 g of fiber and 1.4 g of sugars. For these measures, the Canadian nutritional label has two significant digits, which tells you that Cheerios has 10 percent less fiber and 40 percent more natural and added sugars than indicated by the one-significant-digit American label.

In practice, you often can't be sure how many significant digits there are in a number. When a small air conditioner is billed as having a cooling capacity of 5,000 Btu, you see a number with four digits. But it's hard to believe there are four *significant* digits. Maybe there's only one significant digit, in which case the manufacturer might have rounded off 4,509 Btu to the nearest thousand. Or perhaps there are two significant digits, implying an actual capacity of at least 4,950 Btu, since 4,949 would have been rounded down to 4,900.

The moral is to beware of numbers with zeros at the end. Such numbers were probably rounded, and not necessarily in your favor. Also beware of numbers that may have been converted from other units, because they are often reported with more significant digits than originally existed, thereby giving a false impression of precision. Some years back, a friend of ours who had just read the owner's manual for his new foreign car expressed surprise that the manufacturer reported the fuel tank capacity with such precision: 15.9 gallons instead of an even 16. He would have thought differently about the number had he seen it in its original metric form of 60 liters.

Also be on the lookout for numbers where fractions may have been converted to decimals. When an invoice says, "Labor, 3.5 hours," rarely has time been recorded in six-minute intervals, which is what billing in tenths of hours implies. Likely, 3.5 means $3\frac{1}{2}$. In principle, fractions eliminate such ambiguity, except that we were all taught in grade school to simplify fractions wherever possible, which can limit their effectiveness in conveying information. If a worker's time is billed in 15-minute intervals and he works 3 hours and 30 minutes, that should be recorded as $3\frac{2}{4}$, but it's almost always simplified to $3\frac{1}{2}$ or 3.5, leaving no indication of the billing interval. If this seems like picky stuff, it is, but we're hardly alone in pointing it out.

Rounding off numbers can simplify calculations, but be careful. Computations can magnify the imprecision of rounded numbers, and so it's generally safer to round numbers *after* performing calculations. Approximately how many square feet are in a square mile? A while back we got the exact figure of 5,280 × 5,280 = 27,878,400, which rounds to 28 million. But if we round before multiplying, our answer is much less precise: 5,000 × 5,000 = 25,000,000.

This issue shows up surprisingly often in financial statements. Dell Computer's 2001 annual report says that the company repurchased 68 million shares in that year. Yet when *New York Times* financial writer Floyd Norris totaled the quarterly figures given to him by Dell, he got 69,054,919. A Dell spokesman told Norris that they rounded each quarter's repurchases to the nearest million and then added up the rounded numbers for the annual report. This sloppy quantitative reasoning led Norris to quip that "Someone should buy Dell a computer with a good spreadsheet program."[21]

In deciding how many digits are appropriate for a specific number, context is everything. If you go out to lunch with five friends and are assigned the job of evenly dividing the tab, which with tip totals $85, you don't ask everyone to pitch in $14.17, or even

$14.20. As Miss Manners would surely tell you, they pay $14 each; you pay $15. If you try to be more precise, we'll label you a poor quantitative thinker. Everyone else will just call you cheap.

In other situations, you're making a mistake if you don't measure to as many significant digits as feasible. Before the start of the 2000 Summer Olympics in Sydney, Mike Gibbons, a senior engineer at Swatch, the company responsible for timing events, said that any timing company "worth its salt" could produce equipment capable of timing accurately to less than 1/1,000th of a second. "But," he cautioned, "we're just not allowed to do it."[22] That's right, a third decimal place would violate the rules. As a consequence, Gary Hall Jr. and Anthony Ervin "tied" in the 50-meter freestyle swimming event, when both were timed in 21.98 seconds. Coincidentally, Hall's father, Gary Sr., competed in the race that led to the rule limiting timing intervals to hundredths of seconds. In the 400-meter individual medley final at the 1972 Munich Games, Sweden's Gunnar Larsson and the USA's Tim McKee both finished in 4:31.98, an apparent dead heat. However, when officials decided to look at the third decimal place recorded by the timing equipment, Larsson was awarded the gold medal by two-thousandths of a second. The ensuing controversy led to the two-decimal-place rule, with officials justifying the change on the grounds that it was unfair to decide races by a thousandth of a second when other factors, such as imperfections in pool construction, could potentially affect times by more.

To close the chapter with a bit of editorializing, that kind of "it's too close to call," "it's a shame someone has to lose" thinking has no place in sports. It's all well and good to try to make the playing field as level as possible, but once a competition starts, the playing field has to be accepted as is. The objective in a swimming race is to finish first, not to beat the other competitors by at least 1/100th of a second.

Playing the Percentages

Wyoming has long been considered a Republican state; California has become a Democratic stronghold. Yet there are many more Republicans in California than there are in Wyoming. Toyotas are well regarded for their workmanship; Jaguars have a reputation for unreliability. Yet at any given moment, the average garage is much more likely to be servicing a Toyota than a Jaguar. And Ralph Nader received more votes for president than Abraham Lincoln.

In a world without percentages, chaos prevails.

The absolute numbers implied in the above comparisons are of course begging to be replaced by *relative* numbers, so much so that even neophyte quantitative thinkers are crying out, "Where's the base?" Nader, for example, received 2.8 million votes from a year 2000 electorate that numbered over 100 million, whereas Lincoln managed 1.9 million votes from an 1860 electorate that numbered less than 5 million. So hold the chisels—we don't have to sculpt Nader's likeness into Mount Rushmore just yet.

When all the right numbers are in place, percentage-like calculations can stop arguments before they begin. In 1941, Ted Williams knocked out 185 hits in 456 at-bats, while Joe DiMaggio managed 193 hits in 541 at-bats. But the fans of that era didn't have to enter into parliamentary debate to determine who had the higher average. All they had to do was check that $^{185}\!/_{456}$ = .406, while $^{193}\!/_{541}$ = .357. Williams won the batting title by a comfortable margin.

The key issue here is "common denominator." If the two players had the same number of at-bats in the first place, we wouldn't need any conversion, because whoever had the greater number of hits would have the better average. But common denominators aren't really as common as their name suggests, which is why percentages were adopted in the first place. Percentages essentially force all ratios to have a denominator of 100, at which point these ratios are readily compared.

Lest there be any doubt, and lest your grade school experiences suggest otherwise, percentages exist in order to make our lives easier. When numbers are placed on a scale from 0 to 100, we can see what's puny and what's significant, no matter what the issue is. In the old days, when the Internet served just 2 percent of U.S. households, we knew that enormous growth was in store. And in the even older days, when Ivory Soap's slogan was "99$^{44}\!/_{100}\%$ pure," we knew they had pretty much gone to the limit. Easily understood numbers are particularly welcome in this world, which is why it's just about impossible to read even one section of a newspaper—any newspaper, any day—without coming across percentages. This unique popularity explains why percentages merit a chapter of their own.

To delve momentarily into our elementary school memory banks, we derive percentages from fractions. If your stock portfolio went from $15,000 to $16,000 last year, your absolute gain was $1,000 and your percentage gain was 6.7 percent, because $^{1,000}\!/_{15,000}$ = $^{6.7}\!/_{100}$. The percentage gain provides a basis for comparison. If Warren Buffett's portfolio went from $23 billion to $24 billion during the same period, you can at least pride yourself on having achieved a higher percentage return—6.7 percent versus his 4.3 percent. You can judge for yourself whether the absolute number should in this case carry more weight than the relative number. However, if it's any consolation, if you kept outperforming Buffett at your current pace, you'd draw even in about 647 years.

If the calculation of percentages never held great appeal, we have a couple of pieces of good news, both in the spirit of the "easy way out." The first is that many questions about percentages can be addressed without calculation, by the mere realization that both a numerator and a denominator are involved. For example, which university has a higher percentage of varsity athletes, Princeton or Ohio State? If you distrust the stereotype of Ohio State as being a jock school, you're on the right track. Princeton in fact has 38 varsity sports programs, Ohio State only 30. Couple that with the obvious fact that Ohio State is a much bigger school, and you realize that both the numerator and denominator of our invisible equation are pointing to the same conclusion. Princeton has a higher percentage of varsity athletes than Ohio State does, and it's not even close. Second, and even more basic, most of the percentages we encounter in our lives, whether in the newspaper or in the supermarket, are already calculated for us. Why bother figuring out someone's batting average when you can find it in the sports section?

Unfortunately, that's where our easy way out comes to a screeching halt. Even those grade-schoolers who mastered basic arithmetic discover that real-life percentage figures can be vexingly difficult to interpret. The remainder of this chapter will guide us through the most sinister percentage traps you will ever encounter. Be careful. It's a jungle out there.

The Linearity Trap

The linearity trap lurks in otherwise ordinary-looking percentage calculations, ensnaring millions of unsuspecting Americans every year. Sounds sinister already, but if our goal is to make people truly take notice, there is a better route. After all, in modern society an affliction, condition, or syndrome isn't considered real until it is experienced by a celebrity. So let's turn the clock back to 1989, to an interview conducted by Bryant Gumbel, then of the *Today* show.

The subject of Gumbel's interview was Dr. Charles Hennekens, head of an epidemiological team at Harvard that had just come out with a breakthrough study confirming the long-suspected power of aspirin in thwarting heart attacks. Participants in the Harvard study reduced their incidence of heart attacks by 47 percent by taking one aspirin *every other day*.

Gumbel then stunned Hennekens with his follow-up question—could you reduce the risk of a heart attack by 94 percent if you took an aspirin *every* day?

A major gaffe, obviously. But why?

Medically, Gumbel's question was naïve. Aspirin may be a wonder drug, but if you keep swallowing aspirin tablets nonstop, you will at some point die. It only follows that the improvements cited in the Harvard study cannot continue indefinitely. The every-other-day regimen wasn't an accident. The choice specifically recognized that as dosages of aspirin go up, so do the side effects.

But the arithmetic behind Gumbel's question is even more suspect. Think about it. If you could keep reducing risk *linearly*, after three years your total risk-reduction would be $3 \times 47\% = 141\%$. Which means what, exactly? A risk reduction of 100 percent would mean that your risk of a heart attack had dropped to zero. Does a reduction of 141 percent suggest that you had successfully transferred your risk to someone else?

The immediate lesson here is that you have to be careful in extrapolating percentages beyond the information you have, and unthinking multiplication will almost always get you in trouble. A University of Chicago study of the 1990s revealed that in any given year, about 5 percent of married men have extramarital affairs.[1] Plainly, you can't conclude that over 20 years the list of philanderers would include 100 percent of married men. Even Camille Paglia wouldn't believe that figure.

We'll correct Bryant Gumbel's risk-reduction arithmetic shortly. For now we'll content ourselves with the suggestion that Bryant was unlucky. If the risk reduction from an every-other-day aspirin dose had been 52 percent instead of 47 percent, doubling that number would have pushed him over the 100 percent mark right away and he probably would have caught his error. But now that the rest of us have our celebrity fall guy, we can avoid getting caught by alternative forms of the same trap.

The linearity trap truly blossoms in the realm of portfolio performance. To see why, let's suppose it's 1999 and your investments had done extremely well over the previous two years: Your return for 1997 was 30 percent and in 1998 you outdid yourself—up 40 percent. But how did you do for the two-year period?

Many investors would be content to pat themselves on the back for gaining 70 percent over two years. How else are we supposed to combine 30 and 40? But this simplistic linear step is not only wrong, it is an insult to the power of compound interest. To set things straight, we have to take the vital step of observing that a 30 percent gain is equivalent to multiplying our money by 1.3. Once we decide to think in terms of factors instead of percentages, it is no great stretch to conclude that a 40 percent gain is equivalent to multiplying our money by 1.4. The advantage to this view is that factors are readily combined: Over two years, we have multiplied our money by a factor of $1.3 \times 1.4 = 1.82$. Converting back into percentages, the two-year gain is a tidy 82 percent. You gained 12 percentage points out of thin air.

The reason for the 12-point gap is that the second year's returns were achieved relative to a higher base, because you *made* money the first year. The fact that the bases were different (and will always be different, unless you broke even for a particular period) is why you get an inaccurate result when you add the percentages. And the 12 points that lie between the straight-line calculation and the actual return are not to be sneezed at. The more years you tack on, the bigger your base becomes, and before long the compounded returns blow away the "straight-line" standard, even with more ordinary rates of return. If you earned 8 percent on your money for 20 years, the straight-line number of 160 percent would be pitifully inadequate to account for your soaring wealth, because what you've really done is multiplied your money by a factor of 1.08 to the twentieth power, or 4.66. Which in turn corresponds to a gain of 366 percent. (Sorry, but we have to subtract the 100 percent you started with. The failure to do so is a big trap all by itself. To avoid it, just remember that a doubling is a 100 percent gain and a tripling is a 200 percent gain. The rest should fall nicely into place.)

The Trap of Negative Returns

The multiplication principle we just outlined is trivial, yet at the same time not widely understood. The sad truth is that hundreds of mutual fund managers cannot calculate the returns for their own funds.

If you happen to have your money invested with one of those managers, you might find it especially relevant to know how to calculate *negative* returns. Well, suppose you gained 25 percent in your first year and lost 20 percent in the second. How did you fare overall?

There are two traps with this one. The first is that you don't need to invoke negative numbers to get your answer. Instead we observe that the 20 percent loss in year two is equivalent to multiplying your money by 0.8—a negative return corresponds to a factor less than one. Note that Bryant Gumbel sealed his downfall by missing this step. A 47 percent reduction, whether of heart attack risk, dollars, or mental faculties, means that whatever you had initially has been cut to 0.53 times its former self, because 1.00 − 0.47 = 0.53. If you assume that two such reductions could be linked (true for portfolio performance but not true for the aspirin study), you would obtain your total reduction by multiplying 0.53 by itself to get 0.28. In other words, successive reductions of 53 percent leave you with 28 percent of what you started with, so you'd work backward to conclude that your total reduction was 100% − 28% = 72%.

Getting back to our gain and loss situation, the arithmetic is even friendlier. All told, our two-year factor is $(1.25)(0.8) = 1.0$, which means we broke even. Note that the multiplication in question is automatic when we recognize that $1.25 = \frac{5}{4}$ and $0.8 = \frac{4}{5}$. Simple fractions lead to cancelation, streamlining calculations that in decimal form look pretty gnarly. This is one of many advantages of fractions that go unappreciated in our decimal-obsessed world.

Check the figures again. We seemed to gain more than we lost—25 percent up versus 20 percent down—yet over two years all we did was break even. Counterintuitive, yet easily resolved: Our gain was applied to our original investment, and our loss to a larger amount. As long as we ask the question "Where's the base?" we're less likely to be duped.

This principle of changing bases is connected to a fundamental paradox regarding mutual fund performance. How is it possible for a mutual fund to deliver positive, even market-beating per-

formance over a several year period, only to find that the average
shareholder return during that same period is *negative*?

Actually, the situation isn't all that uncommon. When a fund
exhibits superior performance, it attracts new investors: For example, a $600 million stock fund that goes up 80 percent in a
given year might attract enough publicity to swell to $4 billion
in size by year-end. But suppose that the fund's performance then
hits the skids—as with Internet funds in the bubble-bursting
year of 2000. For these funds, the negative returns of the bust
phase were experienced by many more investors than those fortunate few who enjoyed the superior returns of the boom.

John Bogle, longtime chairman of the Vanguard mutual fund
group, notes that the tendency of investors to flock to hot funds
creates a kind of systematic bias in mutual fund performance statistics:

> Within the fund industry, it is no secret that the conven-
> tional rates of return to measure a fund's performance
> (time-weighted, on a per-share basis), with few excep-
> tions, reflect performance that is significantly higher, and
> in many cases radically higher, than the returns actually
> earned by its shareholders (dollar-weighted, on the basis
> of total net assets).[2]

And Bogle provides an example demonstrating that the juxtaposition of positive time-weighted returns and negative dollar-
weighted returns is not simply an artifact of the late 1990s
Internet bubble. Bogle notes that "the fund with the highest
(conventionally measured) return in the entire industry—annually, about 20 percent per share—in the decade ended July 31,
1996, had a dollar-weighted return of –4 percent during the same
period. There is a difference and investors should be aware of it."

When the chairman of a mutual fund company gives you warnings about the calculations made by his own industry, it doesn't hurt to listen.

Percents of Percents

If you dislike having the wool pulled over your eyes, this final example of performance calculation will have special appeal. Some years ago the brokerage firm Smith Barney (before it was acquired by Salomon Brothers to form the present-day behemoth Salomon Smith Barney) released its annual list of favorite stocks, in which each industry analyst would pick out his or her single best idea for the upcoming year. This list did extremely well year after year and understandably became a source of corporate pride. But one year they went too far.

Here's the setup: Suppose the S&P 500 was up 30 percent for a given year, while your list of stocks was up 44 percent. What would you say? That you beat the S&P by 14 percent? Well, technically, you beat the index by 14 percentage points, and therein lies an opportunity for some mischief. When Smith Barney faced those same numbers back in 1991, someone in the editorial department apparently decided to create a spin. The headline in their *Portfolio Strategist* publication read that the favorite stock list had outperformed the index by 48 percent.[3] Well, it did, literally: the number 44 is 48 percent greater than the number 30. (Actually, it's 47 percent greater—there were rounding issues here as well.) But that's not the way investors talk. If the Fed funds rate is 4 percent and you hear that it was cut by "one percent," you know that it is now 3 percent, not 3.96 percent, because the Federal Reserve doesn't bother with cuts of 0.04 percentage points. Just as surely, when you speak of beating the market by 48 per-

cent, it sounds as if you were up 78 percent (30 + 48). The confusion between percent and percentage points had the effect of
adding 34 percentage points to Smith Barney's performance.

A different conundrum involving percents of percents showed
up in a significant national debate that picked up steam in the
late 1990s. The subject was the partial privatization of Social
Security, under which a certain percentage of a worker's Social Security contributions would be diverted into a private account.
The worker would then have the freedom to invest the money in
that account in the stock market. In that way, Social Security
would become a bona fide investment vehicle and not simply a social insurance program that transfers wealth from one generation
to another. Depending on the side to which you were listening,
you heard this described as either a wonderfully progressive development or a scurrilous attempt to place workers' hard-earned
savings at risk. Without delving into the politics of the situation,
surely it was impossible to evaluate the plan without understanding one basic issue: How partial is "partial"?

That's where things got tricky. Remember, Social Security had
over the years earned the designation "the third rail of American
politics." You touch it, you die. With that backdrop in mind,
here's what one promoter of partial privatization, Democratic representative Charles Stenholm of Texas, had to say of legislation he
introduced: "We have put our name on a proposal that uses private markets for two percent of what's now going into Social Security."[4] Just 2 percent? Sounds too trivial to fight, doesn't it?
Well, that was presumably the method in the madness.

But if you really want to answer the question "How partial is
partial?" you have to understand that the average worker's Social
Security contribution amounted to 12.4 percent of wages (only
half of which is paid by the worker, the other half being paid by
the employer). That "2 percent" was in fact 2 percentage points of

those 12.4 percentage points, or about one-sixth of the total Social Security contribution. Obviously this is a much bigger number, and the correct one at that. Stenholm could have easily replaced the words "two percent" with "one-sixth" in his public statement, but the bizarre nature of this particular debate was that the people in favor of devoting a percentage of the payroll tax into private accounts were trying to make the number seem as small as possible, a problem that endured throughout the 2000 presidential campaign. Like Nancy Reagan's "itty-bitty gun" or Bill Clinton's "I didn't inhale," this one left all of us dazed and confused.

We've got one more variation on this theme. Everyone knows that college admissions have gotten tougher and tougher over the years. Consider that in 1990, the University of Pennsylvania accepted over 42 percent of its applicants, but by 2000 that figure had dropped to just 23 percent. And when the Ivy League gets more competitive, so does the entire nation, on the trickle-down theory that when the most prestigious schools rebuff thousands of qualified candidates, those same students are forced to broaden their applications, thereby crowding out the next tier, and so on. It turned out that for the year 2000, the acceptance rates at Harvard, Princeton, and Yale were 11.1 percent, 12.4 percent, and 16.2 percent, respectively, each figure having dropped significantly during the prior ten years. Those are remarkably low numbers, all the more so given the self-filtering nature of the applicant pool to the nation's finest universities. But just how close are those figures?

Certainly 11.1 percent and 16.2 percent appear close together from our customary vantage point—on a scale that stretches from 0 percent to 100 percent, they are separated by only five points. But now look at the underlying numbers. Conveniently, Harvard and Yale accepted almost the exact same number of students in

2000: 2,082 for Harvard versus 2,084 for Yale. So suppose you asked the pivotal question from the standpoint of the admissions committee: "How many applicants must we look at to generate those final numbers?" If you plug in the acceptance rates, a chasm suddenly emerges. Whereas Yale chose its class from an applicant pool of 12,887, Harvard's applicant pool numbered 18,693. From this perspective, the two numbers aren't even close, and indeed, 18,693 is 45 percent bigger than 12,887. But this disparity isn't a surprise, because the acceptance rates of 11.1 and 16.2, viewed as numbers instead of percentages, differ by the same percentage (16.2 is 45 percent more, rounded off, than 11.1).

When Percentages Aren't the Answer

One corollary of the above trap is that percentages aren't necessarily the correct measuring tool, even if in theory they can be calculated. This principle was in full force when the 2000 election finally ended. When the Supreme Court gave its 5–4 decision in *Bush v. Gore*, there was of course widespread disagreement about the fairness of the conclusion, yet no one on either side took the one-in-nine majority to mean that Bush had won by 11 percentage points!

The reason why no one spoke in terms of percentages is that you can't extrapolate from nine Supreme Court justices: They are what they are. Percentage designations are reserved for bigger sample spaces. Obvious? Of course, but violations of this sort crop up all the time.

Professional pollsters have received all sorts of criticism in recent years, a result of the dizzying output of pre-election polls combined with the fact that so many of these polls have proved so wrong on Election Day. *Poll*ution, you could call it. Polling is of course an inexact science, one that we will cover at greater length in chapter 8. But by giving us the results in percentage terms, polls suggest a sense of permanence that they don't really deserve. Set aside the plus-or-minus 3 percentage-point caveat that comes along for the ride; when we hear that one candidate is ahead 46 percent to 38 percent, we in effect allow those figures, which may have been created on the basis of 500 or so interviews, to cast a far wider net. So when the next poll comes in at 43 percent to 40 percent—possibly with the lead reversed—we are stunned. If we saw the absolute figures behind the polls, we'd be better equipped to deal with their volatility, but you have to bring out your magnifying glass for that.

As if the 2000 election didn't give us enough material, the tax-cut battle of that campaign gave us a completely different instance of percentages vying with absolute numbers. The battle began when Republicans promoted an across-the-board tax cut that was consummately fair in the sense that everybody, whether wealthy or middle-class, would receive the same percentage cut. How could you possibly argue with that?

The Democratic side had little problem coming up with a counterargument. Their approach was to view the issue in absolute terms. If you cut everyone's tax bill by the same percentage, a high earner will necessarily get a bigger tax cut in absolute terms than a low earner. The Republican tax cut was apparently fair in percentage terms, but it was undeniably an initiative that would funnel most of its dollars into the hands of those who, by definition, were rich to begin with.

Our purpose in raising this issue isn't to take sides. Rather, it is to point out that percentages and absolutes don't always go hand in hand, even when one is built from the other. Years ago the State of Connecticut legislature decided that it would help towns meet the cost of new school construction by matching the town's cost dollar for dollar. In other words, the state would foot 50 percent of the statewide school construction bill. In practice, however, they (and many other states) realized that such a "matching fund" system was in effect a pool of dollars that was distributed disproportionately to wealthier communities, the very places that least needed the assistance. As a result, Connecticut shelved the 50-50 system in favor of a "sliding scale" system, in which the state would reimburse cities and towns at a rate as low as 20 percent (for wealthy communities such as Greenwich and Darien) and as high as 80 percent (for struggling cities such as Bridgeport or farming communities such as Putnam). As the saying goes, what's fair is fair. But defining fairness with a single percentage can be asking for trouble.

The 0-to-100 Trap

Although the tidiness of the 0-to-100 scale is a great strength of the percentage system, many percentage figures turn out to dwell in a much more limited range. For example, presidential approval ratings theoretically occupy the entire 0–100 range, but the lowest figure ever recorded was Harry Truman's popularity rating of 23 percent, achieved (if you want to call it an achievement) in the fall of 1951. It is reasonable to treat the 23 percent figure as rock bottom, even more reasonable when you consider that Richard Nixon achieved an approval rating of 24 percent immediately prior to his resignation.

In the same spirit, a survey result that comes out 50 percent "yes," 30 percent "no," and 20 percent "undecided" appears to give the "yes" side a 20 percentage-point advantage. But you could argue that the real base here is 80, not 100. Of those who've made up their minds, 62.5 percent said "yes" and 37.5 percent said "no," a difference of 25 percentage points.

Some percentage figures can be huge without getting anywhere near the upper limit of the 0–100 range. As long as we're talking about presidents, we see that to a presidential candidate, 65 percent is a de facto upper limit; even landslide winners such as FDR and Nixon never garnered that share of the popular vote. And "landslide" may be the appropriate term, because the constraints in real-life percentages are similar to constraints in the slope of the land. We know that a zero-degree slope represents flat land and a 90-degree slope represents a sheer cliff, but among the roads we'd actually drive on, even a 20-degree slope would be extraordinary, despite the fact that the 20 figure looks innocuous at first glance. And no one but the most expert skiers would tackle a 45-degree slope.

We note with some amusement that many slopes are measured in "grade" rather than degrees: a 45-degree slope, because it represents one unit across for every unit down, is said to have a grade of 100 percent. In practice, slopes seldom exceed 45 degrees, which is to say that grades seldom exceed 100 percent. But we're back to the old 0–100 scale again, such is its power.

Because the 0–100 scale is so widely accepted, we can get into real trouble if we go into triple figures. It does happen, of course. There are no limits on the percentage gain a stock can produce, for example, although your relationship with your own broker might make you feel differently. But unless you bought on margin, you at least know that the stock can't go down by more than 100 percent. And you can't actually give 110 percent on the playing field, no matter what the coach—or whoever wrote the deodorant commercial—happens to say. In fact, if you could give 110 percent, it would be a pretty measly effort, given that such numbers as 180 percent or even 900 percent would be at your disposal.

Strange things happen with percentages of three digits or more, and they are often worthy of suspicion. Consider the "Movers and Shakers" list on Amazon.com. The folks at Amazon decided that in addition to giving every book on their list a sales rank, they would alert people as to which books were showing the greatest upward movement in terms of that sales rank. (The M&S list was concocted in the middle of Harry Potter mania, so perhaps the Amazon people were tiring of having Harry's books the only ones that appeared on the home page.) One day in March 2001, the #1 Mover and Shaker was none other than *Green Eggs and Ham*, by Dr. Seuss. Alongside the book's name was the numerical phrase "+4,310%."

But what could that figure possibly mean? It turned out that for some reason or another, the book had gone from 441 to 10 on the list, and 441 is 4,310 percent larger than 10, which is how Amazon arrived at their figure. In other words, in order to generate a big number, the conventional roles of numerator and denominator were interchanged. (It got worse. In July of that same year, following the death of *Washington Post* publisher Katharine Graham, her memoir *Personal History* moved from 501 to 1, a "gain" of 50,000%!)

Somehow we are reminded of times when people refer to a quantity as "twice as small" as another, or to a price as "four times as cheap." For example, in describing the artificial organisms of nanotechnology—notably the microparticles that behaved so badly in his book, *Prey*—author Michael Crichton wrote, "Such man-made machines would be 1000 times smaller than the diameter of a human hair."[5] Although such phrases are familiar enough that we understand them, they are, strictly speaking, nonsensical. We can say that a 200-pound person is "twice as heavy" as a 100 pounder because $2 \times 100 = 200$; or we can say that the lighter person is "half as heavy" because $\frac{1}{2} \times 200 = 100$. But where's the equation for "twice as light"? It doesn't exist, because the arithmetic isn't in the right direction. If you run the 100-meter dash in 20 seconds, you wouldn't give a world-class sprinter like Maurice Greene much of a match, because he comes in at under 10 seconds. Would you then say that you are twice as slow as he is?

In Amazon's case, their M&S ratings had to be amended when they realized that an author whose work was struggling with a seven-digit sales rank could propel his book into Mover and Shaker status by simply buying a handful of copies, because that might be enough to get a rank of somewhere in the thousands. Wanting to discourage such hokum, Amazon decided to restrict the M&S list to books already ranked between 1 and 1,000. The problem is that the real movers and shakers—the books that had gone from oblivion to prominence overnight, whether because of a *New York Times* book review or an appearance on *Oprah*—were sure to be eliminated. Unless we truly believed that Dr. Seuss was the hottest thing going in March 2001.

There's an important general lesson in the M&S story. As we emphasized in the previous chapter, most numbers count or measure things. Rankings, whether of sales, city populations, or figure skaters, are different. Like page numbers and the number at a deli showing which customer is being served, rankings belong to a special class of numbers called ordinal numbers. Ordinal numbers, as you might guess, simply indicate order and as such aren't really numbers, in the sense that order can just as well be indicated by the letters of the alphabet. (As are the columns on a spreadsheet.) Now for the lesson: Be wary of doing arithmetic on ordinal numbers. Suppose that a book's sales volume increased by 10 percent from one week to the next, moving it from 24th to 6th on a ranking list. To say the book advanced by 300 percent (the percentage by which 24 exceeds 6) is next to meaningless, precisely because the calculation is based on ordinal numbers, not the quantity of books sold. The point would be obvious if the rankings used letters instead of numbers—no one would attempt to calculate the percentage gain represented by moving from X to F.

A similar lesson applies to percentiles, which express rank in percentage terms. When a schoolgirl scores in the 90th percentile on a standardized test, it means she has done better than 90 percent of the kids taking the exam. That's all the score means. It's tempting to assume that 90th percentile is a much better performance than 45th percentile—the numbers make 90th percentile seem twice as good as 45th percentile. But percentiles don't tell us that. As a form of ordinal numbers, percentiles simply report rank and not underlying values. Depending on the circumstances, a 90th percentile performance could be dramatically superior to one in the 45th percentile, or there could be little difference between the two performances. In some cases, 90th percentile might signify an excellent outcome, while in other situations 90th percentile might be a disappointing result.

In ranking mutual funds, Morningstar awards a fund one to five stars based on a percentile ranking of the fund's historical performance (after adjusting for risk and sales charges) against similar funds. Funds whose past performance places them in the top 10 percent in their fund category (90th percentile or better) are awarded five stars. The next 22.5 percent of funds in a category receive four stars; the middle 35 percent earn three stars; the next 22.5 percent receive two stars; and the bottom 10 percent get one star. Because the star ratings represent percentile ranges, they don't indicate the actual monetary performance of funds, either in absolute terms or relative to their peers. In some fund categories, such as short-term government bond funds, the difference in returns on a one-star fund and a five-star fund might amount to no more than a percentage point a year over a five-year period. Within other categories, such as technology stock funds, the gap in five-year returns between a one-star and five-star fund could easily exceed 20 percentage points annually. Moreover, a one-star fund in a strong category will often outperform a five-star fund in a weak category. As this was being written in September 2002, the five-star Scudder Latin America S Fund had lost a cumulative 37 percent over the previous five years, while the one-star Smith Barney Government Securities A Fund had gained 35 percent over the same period.

You Can't Get There from Here

All men are mortal. Socrates was mortal. Therefore, all men are Socrates.

Spotting the logical flaw in Woody Allen's version of the classic syllogism is easy. Yet when percentages are involved, people constantly reach conclusions that are just as unfounded as the deduction that all men are Socrates. The root cause is that many statements involving percentages look more informative than they really are. If a dollar buys four bananas, we can deduce that a banana costs 25 cents. But the fact that 40 percent of minivans are Chryslers doesn't tell us what percentage of Chryslers are minivans. In theory, that percentage could be anything above zero up to 100. And it comes from dividing the number of Chrysler minivans by the total number of Chryslers, not by the total number of minivans, which is where the 40 percent came from.

Perhaps the example of Bill O'Reilly can scare you straight on this percentage trap. In a 2002 interview with National Organization for Women president Kim Gandy, the host of *The O'Reilly Factor* asserted that "58 percent of single-mom homes are on welfare, 58 percent." Although Gandy claimed the figure was "not even close," O'Reilly insisted it was correct. "I'm not misreading it. It's 58 percent. That's what it is from the federal government. If you choose not to believe it, fine." However, O'Reilly subsequently backtracked, offering a quite different statistic: "Fifty-two percent of families receiving public assistance are headed by a single mother, 52 percent."

At first glance, it appears that O'Reilly was simply off by six percentage points, but look again. In addition to the six-point slip of the tongue, he subconsciously equated two very different percentages. We suspect that O'Reilly found out that single moms head a majority percentage of families receiving public assistance, and then unknowingly made the bogus inference that a majority percentage of single-mother families receive public assistance. Such a mistake is more understandable than forgivable, but the main point is that you can't compute the latter percentage from the former. And it turns out that the percentage of single-mother families receiving public assistance, while significant, is well below 50 percent. When Gandy returned on a subsequent broadcast, O'Reilly ceded some more ground and fell into a different percentage trap in the process. "What we have ascertained is 33 percent of all single mothers receive food stamps. Their families receive food stamps. Another 14 percent receive federal welfare, but the real stat—" At which point Gandy interrupted O'Reilly by saying "Not *another* 14 percent," knowing that the population of food stamp recipients of course overlaps with the population of welfare recipients.

Percentages, Policy, and Pareto

Our final examples of percentage nuances will confirm the importance of asking "Where's the base?" whenever percentages are involved. Ordinary looking numbers can produce extraordinary results unless we've got the wrinkles down pat.

Suppose that as the new century dawned, you wanted to find a way to give some money to your favorite nephew. The exact amount is irrelevant, as long as it's enough to put you in a suitably high tax bracket: We're feeling generous, so we'll use $1,000,000. We'll even assume that your prior generosity had placed you in the top gift and estate tax bracket, which in 2000 had a marginal rate of 55 percent. From a taxation standpoint it didn't seem to matter whether the gift was made immediately or instead passed on as part of your estate; either way the tax rate was 55 percent. Leaving the issue of timing aside (obviously it might make a tremendous difference *when* the money was transferred, especially to your nephew), we'll just address the simple question: How much would it cost you to deliver the $1,000,000?

That's where we learn that not all 55 percents are created equal. If the $1,000,000 was in the form of a gift, the IRS would tack on what amounts to a 55 percent surcharge. You write the government a check for $550,000 and the transaction is complete. Your net cost: $1,550,000.

However, if the favorite nephew was to receive $1,000,000 from your estate, your initial allotment would have to be $2,222,222—that's the number which, when you take away 55 percent, leaves you with $1 million. As long as we're dealing in percentages, note that this number is 43 percent greater than the $1,550,000 figure used in the gift tax calculation. The huge gulf arises because the two 55 percent taxes are identical only in number; they represent two completely different approaches to taxing transfers. When we said that the best quantitative thinkers "never trust numbers," we weren't kidding. Perhaps we should have added that they do it out of necessity.

In the early 1980s, when welfare dependency was a prominent political issue, Harvard researchers Mary Jo Bane and David Ellwood discovered that at any given time, about 60 percent of the poor were in the midst of poverty spells that would eventually last eight years or more.[6] In an apparent contradiction, Bane and Ellwood also found that most poverty spells are relatively short. Specifically, about 40 percent ended within a year and two-thirds were over within three years. Only 15 percent of poverty spells lasted longer than eight years.

What gives?

Bane and Ellwood offered an explanation by way of analogy:

> Consider the situation in a typical hospital. Most of the persons admitted in any year will require only a very short spell of hospitalization. But a few of the newly admitted patients are chronically ill and will have extended stays in the hospital. If we ask what proportion of all admissions are people who are chronically ill, the answer is relatively few. On the other hand, if we ask what fraction of the hospital's beds at any one time are occupied by the chronically ill, the answer is much larger. The reason is simple. Although the chronically ill account for only a small fraction of all admissions, because they stay so long they end up being a sizable part of the population in the hospital and they consume a sizable chunk of the hospital's beds and resources.
>
> The same basic lesson applies to poverty. Only a small fraction of those who enter poverty in any given year will be chronically poor. But people who will have long spells of poverty represent a sizable portion of the group we label "the poor" at any one point in time.

The implication for welfare policy was that both the 60 percent and 15 percent figures could be misleading. If you took a snapshot of the welfare rolls, it looked as though 60 percent of those receiving welfare were either long-term dependents or on their way to long-term dependency. Arguably, that makes welfare look like more of a trap than it was. On the other hand, the fact that only 15 percent of those who went on welfare became long-term recipients seems to understate the magnitude of the problem of long-term dependency, which accounted for most of the cost of welfare. (The comparable figures are quite different today because most poverty assistance is now subject to time limits. In 1996, Aid to Families with Dependent Children [AFDC] was replaced by Temporary Assistance for Needy Families [TANF].)

Note that Pareto's Law is at the heart of this paradox of percentages. The bulk of welfare costs were spent on that group of chronically poor who, though relatively small in number, accounted for a disproportionate share of all welfare cases. This time around, the dichotomy was 60/15 rather than the conventional 80/20, but the principle is the same—so, for that matter, is the ratio.

It is fitting that we invoked Pareto's Law to help unravel a paradox of percentages, because our final example involves the law itself. Law school grads who took the New York bar exam in July 2002 heard that the pass rate for the previous July was 70 percent (6,475 of 9,194). Not a bad showing, but that number, even if repeated, didn't really apply to any of the hopefuls for 2002. The Pareto element is that a significant chunk of any year's test-takers are taking the test for a second (or third, or fourth) time. The pass rate for first-timers in July 2001 was a whopping 80 percent (4,089 of 5,136), from which we can subtract out to conclude that the pass rate for the non–first-time takers was 2,368 of 4,058, or 58 percent. No matter which group you belonged to, that single-point estimate of 70 percent misstated your a priori chances by at least 10 percentage points.

Epilogue

As widespread as percentages are, we can cite one instance where percentages played a role in that rarest of twentieth-century phenomena—a decline in quantification. Recall that in 1941, Ted Williams won the batting title with an average of .406, a mark that hasn't been matched since, and a mark 49 points ahead of the .357 that DiMaggio posted that same year. Yet DiMaggio won the Most Valuable Player award that year, buoyed by his record 56-game hitting streak. Never mind the rash of favorable official scorer decisions that kept the streak alive, and never mind the fact that DiMaggio's batting average during his streak—.409—was just three points higher than the level Williams sustained for the entire season. If you sat in the Fenway bleachers when the Yankees were in town, you could still provoke a fight about who most deserved the award.

What few people remember is that the MVP award dates back to something called the Chalmers Award. Back in 1910, Chalmers was an automobile manufacturer in search of a little publicity, and as the baseball season began the company announced it would give one of its shiny new roadsters to the batting champion of each league. At season's end the American League champion appeared to be Napoleon Lajoie, who led Ty Cobb by something on the order of .384 to .383. That's what numbers are for—to settle disputes once and for all, right? But the record-keeping of the era was both manual and sloppy, making those numbers especially untrustworthy. The ultimate insult was that Lajoie's razor-thin margin hinged on his bounty from a final-day doubleheader, in which he beat out no fewer than six bunts down the third-base line, all this against a St. Louis Browns third baseman who remained stationed back near the outfield grass on orders from his manager. Chalmers didn't exactly have this sort of "measurement changes behavior" controversy in mind (the Browns' manager was later fired), so Cobb and Lajoie each received a car. The company then revamped the award for the following season, declaring that they would rely on a panel of sportswriters to determine the individual most worthy of recognition. Chalmers, alas, would not endure. But the MVP award, for all its imperfections, lives on and on.

5

Gaining Perspective

In the early 1950s, when *What's My Line?* panelist Steve Allen wanted to learn more about the product made by the mystery guest, he launched the now-classic query, "Is it bigger than a breadbox?" A couple of decades later, when Johnny Carson had hit his stride on *The Tonight Show*, he couldn't begin a joke with a line such as "It was hot in Los Angeles today" without someone in the studio audience yelling out, "How hot *was* it?" Well, when a number arrives at our doorstep without any obvious benchmarks, our curiosity should at least be up to television standards. We owe it to ourselves to ask questions such as, "Is that a lot?" or "Should we be impressed?" In other words, you can't play the numerical version of 20 Questions without at some point asking, "Is that a big number?"

If you ask a bunch of schoolchildren for examples of big numbers, they'll point to whoppers like "a million," "a trillion," or even "a zillion." Only years later will these kids learn that numbers aren't big or small without some sort of context. One million can be a puny number if it represents miles in the solar system or even dollars in a company's stock market capitalization. By the same token, the number 0.18 percent is tiny in most circles but gigantic if it shows up on a Breathalyzer test. Perspective is everything.

The importance of perspective isn't a new concept for most people. It isn't even a new concept for this book. We have already written thousands of words on percentages, where the whole point is that the relative size of numbers can be more important than their absolute magnitude. And within the measurement chapter, one of our early themes was that numbers delivered without easily understood units aren't much use. But numbers without perspective are worse than useless; they are downright dangerous.

Our aim in this chapter is to help you become a Babylonian. Whatever image that evokes, it's probably not what we have in mind. In a famous lecture series, the renowned physicist Richard Feynman contrasted what he termed Babylonian and Euclidean methods of reasoning. Mathematicians, said Feynman, typically follow the Euclidean approach, attempting to logically derive truths, in step-by-step fashion, from first principles or axioms. The best theoretical physicists, by contrast, employ the Babylonian method, which flexibly assumes that theories can come from different conceptualizations of the same phenomena. Feynman suggested that "every theoretical physicist who is any good knows six or seven different theoretical representations for exactly the same physics. He knows that they are all equivalent . . . but he keeps them in his head, hoping that they will give him different ideas."[1]

Good quantitative thinkers are Babylonians. They understand that quantities can be measured and expressed in many different ways and that looking at something from multiple viewpoints enhances perspective and fosters creative thinking. Put another way, if you are trying to understand the terrain in an area, you're better off studying several maps instead of just one. How much did the federal government spend on national defense in fiscal 2001? The answer is $309 billion (excluding veterans benefits and services). But defense spending in 2001 was also $1,079 per American, 17 percent of federal outlays, 3 percent of the country's GDP, 4.8 percent higher than in the previous year in nominal dollars, and 2.5 percent higher than in 2000 when adjusted for inflation. For a skilled quantitative thinker, defense spending is all of these numbers. Achieving that type of perspective is what this chapter is all about.

Orders of Magnitude and the Principle of Proximity

In the 1987 movie *Broadcast News*, budding anchorman William Hurt made Holly Hunter's eyes roll when he cited a flawed Defense Department initiative that had already cost the government $5 million. "That's *billion*," she fired back.

Well, of course it's billion, when you think about it. If residential homes can be priced in the millions, a U.S. Army program to develop a state-of-the-art "General Stillwell" tank fleet just can't be in the same ballpark. A million, a billion, and all the other illions may *look* similar, but they represent entirely different orders of magnitude. A million is a thousand thousands. A billion is a thousand millions. A trillion is a thousand billions, or, if you prefer, a million millions. As for a zillion, that's just a catchall word for a really big number.

Unfortunately, these different orders of magnitude often aren't appreciated for what they are. And it's not only dumb movie characters who get tripped up. In *Contact*, the astrophysicist portrayed by Jodie Foster made the following argument about the likelihood of extraterrestrial life:

> Out there, just in our galaxy alone, there are 400 billion stars. If only one out of a million of those had planets, and if just one out of a million of those had life, and if just

one out of a million of those had intelligent life, there
would be literally millions of civilizations out there.

"Wrong," notes Neil deGrasse Tyson, director of the Hayden
Planetarium and a real-life astrophysicist. "According to her
numbers, that leaves 0.0000004 planets with intelligent life on
them, which is a figure somewhat lower than 'millions.' No doubt
'one in a million' sounds better on-screen than 'one in ten,' but
you can't fake math."[2]

The underlying issue here is that as human beings, we have
trouble coming to grips with all sorts of big numbers. Numbers
are without bound, but that is not true of life. Life is finite. Resources are finite. Quite unconsciously, people analyze numbers
according to what we will call the principle of proximity: People
feel much more comfortable with numbers that are within the
spectrum of their daily lives, and numbers outside that range frequently end up getting distorted.

What's the difference between a hundred million and a hundred billion? A factor of one thousand, right? But what's the difference between a hundred million dollars and a hundred billion
dollars? That's a completely different question, with a completely
different answer. In life as most of us know it, there is almost no
difference between the two amounts because both fall into the "all
the money you'd ever need" category. As a case in point, when
Mark McGwire announced his retirement from baseball at the
conclusion of the 2001 season, he also walked away from a two-
year, $30 million contract he had agreed to but never signed. For
most people, that $30 million would have been the decisive ele-
ment in the decision, but McGwire had the luxury of knowing he
didn't need the money.

Have you ever heard someone refer to a number as "almost infinite"? Strictly speaking, the phrase is ludicrous: Any number,
however large, is finite, whereas infinity represents the concept of

boundlessness. But a number can be *functionally* infinite, if it represents an amount of time, distance, money, or whatever, that seems beyond human proportions. Mark McGwire's lifetime earnings would certainly qualify.

William Hurt's character may have been quantitatively unskilled, but the English language set him up for such a gaffe. Not only do million and billion differ by only a single letter, but you have to wonder if it's a mistake to introduce different terms every time you add three zeros. The Japanese evidently think so. Whereas English comes up with new terms for big numbers at three-digit intervals—millions, then billions, then trillions—Japanese terms change every four digits. There is a word for 10,000, *man*, and everything larger is expressed in *man* until you get to 100,000,000, *oku*, which is used until you reach 1,000,000,000,000, or *chou*. *Man*, *oku*, and *chou* cover a wider range than million, billion, and trillion—and they don't rhyme.

If English presents challenges for those wrestling with big numbers, the language of mathematics can be even more daunting. The basic difficulty is that mathematical notation for big numbers revolves around exponents. Exponents were designed to make our lives easier—they convert multiplication into addition, as in $10^4 \times 10^5 = 10^9$—but schoolchildren react to them with an instinctive hatred, as if they were cauliflower or broccoli.

When dealing with big numbers, exponents show up even when they aren't invited. If you punch out $123,456,789 \times 123,456,789$ on a calculator, your answer will consist of the number 1.524 together with a 16 set off to the right. The 16 is an exponent that denotes an order of magnitude, as in 1.524×10^{16}, but no one is especially happy to see it. What the calculator doesn't understand is that where exponents are concerned, the fears, misunderstandings, and errors of our school days get carried straight into adulthood.

To take a simple example, 1,000 is a four-digit number, so it's easy to slip up and identify it with 10^4. But the exponent in a power of ten refers to the number of zeros following the one, not the total number of digits in the number, so 1,000 must equal 10^3 instead. Now, being off by one is hardly a crime. If you're printing pages 3–9 of a document, a mental subtraction registers six pages, but, barring a paper jam, seven pages will emerge from the printer. Similarly, if you're looking at check number 2996 in your checkbook and you know that check 3000 is the last one, you have five checks remaining, not four. And so on. No great harm done, but when you're off by one in an *exponent*, you're by definition off by an order of magnitude. In a decimal world, an order of magnitude is a factor of ten. A miss is truly as good as a mile.

Numbers on a National Scale

All too often, numbers that we are conditioned to view as big numbers effectively shrink when we broaden our context. Continually striving to enhance our perspective on seemingly large numbers is essential if we are ever to grasp issues that confront a country as big as the United States.

Ethanol is a gasoline additive/substitute that in the United States is primarily derived from corn, but can also be produced from sugar cane, sorghum, wheat, and other renewable sources. Ethanol burns cleaner than fossil fuels. Sounds good so far: Anything that could alleviate America's chronic reliance on imported oil and simultaneously aid the environment has to be considered a good thing, right? From a standing start in the late 1970s, the total ethanol output of the United States reached 1.8 billion gallons in 2001.

But is that a big number?

Schoolchildren would nod excitedly, but another number from 2001 puts things in perspective: that year, Americans consumed over 130 billion gallons of gasoline.³ Right away the ethanol number looks pretty small, so perhaps we should shift gears and ask if it isn't still possible for ethanol to be the fuel of the future. Alas, there is a fundamental factor that works against ethanol's potential, and we're not talking about the politics of subsidies, the difficulties of achieving nationwide distribution, the power of the oil lobby, or any of the other practical hurdles faced by a gasoline alternative. The factor to which we are referring, believe it or not, is that America isn't big enough.

The key statistic, though meaningless for the moment, is that in 2000, U.S. farmers harvested 72.7 million acres of cornfields. On average each acre yielded 137 bushels of corn, for a total of just under 10 billion bushels.⁴ A bushel of corn can be converted into just over 2½ gallons of ethanol.⁵ Multiplying a shade less than 10 by a touch more than 2½ gives us 25 billion gallons of ethanol. But there's one more step here: When burned, a gallon of ethanol produces only two-thirds as much energy as a gallon of gasoline, so having 25 billion gallons of ethanol is like having 16.7 billion gallons of gas. And with that we get our punch line: Even if we devoted 72.7 million acres (nearly one-fourth of our farmland) to ethanol production—leaving ourselves and the animals we feed on awfully hungry—we'd still have only enough ethanol to replace 13 percent of our gasoline consumption.

Our calculating leads us to the realization that annual U.S. ethanol output, although it amounts to billions of gallons, is not really a big number, and probably never will be if our aim is to supplant a large share of our fossil fuel consumption. All the more so when you consider that it takes a fair bit of energy to grow corn, harvest it, and distill it into ethanol; Department of Agriculture researchers estimate that ethanol yields just 34 percent more energy than is used in its production.⁶ And most of the energy used to create ethanol comes from fossil fuels.

Note that the final figure for how much gasoline could theoretically be saved if we earmarked all of our corn production for ethanol depended on a couple of key numbers that aren't exactly considered common knowledge: the total acreage of corn harvested, the yield per acre, the volume of ethanol obtained per bushel of corn, and the relative energy content of ethanol and gasoline. You didn't think we *knew* those figures, did you? Of course we didn't know them; we searched the Internet. The availability of key numbers from a quick and easy web search is one of the unsung benefits of the Internet age. Most reference numbers you need are just minutes away.

The most lasting imprint of our calculation, however, is the importance of a word we haven't even used yet, and that word is "dimension." The calculation demonstrated that the world's energy producers can easily pump more oil and gas than can be "matched" by the agricultural landscape of the United States. This paradox arises because the former is a three-dimensional construction and the latter is only two, a dimensional component that is at the root of many problems in perspective.

It's not immediately obvious, for example, that we use less water in taking a shower than in taking a bath. But to fill a bathtub even halfway with a running shower would take more time than we would ever care to spend with water pouring down on us, all the more so with the advent of water-efficient showerheads. The shower is in some sense a two-dimensional construction trying to fill a three-dimensional space. Of course, you could say the same thing about the process of drawing a bath, but then we'd be back to the point that giving yourself enough water in which to take a bath requires leaving the faucet on for much longer than our intuition would tell us.

Housepainting, whether inside or out, provides another nice demonstration of the extra dimension at work. Unless you're working on a textured or highly porous surface, the staying power of a gallon of paint is prodigious, because each layer is microscopically thin. The recycling process also involves some dimensional aspects. At first it seems strange that the bottle and can recycling machines at the local supermarket insist on crushing—or, in the case of glass bottles, shattering—whatever they take in. But with a limited amount of storage space available, that crushing is essential in producing three dimensions' worth of recyclable materials, rather than two dimensions of recyclables and one dimension of air. That same principle extends to the local dump, where containers for recycling corrugated cardboard are designed with a vexingly small opening, leading to the comical sight of town residents stamping furiously on their cardboard boxes until they lie flat enough to push through.

To take the concept of a town dump to a more global level, suppose we wanted to dig a hole to accommodate all of mankind's garbage for the twenty-first century. Let's say the hole is 500 feet deep. Based on the latest estimates, how much territory do you think our garbage pit would take up? Our instincts tell us that 500 feet is a small number, so that the space would therefore have to be enormous, but the actual number turns out to be on the order of just 10 square miles. The presence of the third dimension changes everything, and in fact 500 feet is a pretty big number in this context, once you consider the real-life constraints to actually digging out 10 square miles to such a depth. Yes, if we filled our hole with paint, we could probably cover the entire planet several times over, but that's a calculation we leave you to ponder.

Pareto's Law and Political Symbolism

There is a peculiar arithmetic associated with orders of magnitude. Within that arithmetic you will actually see rules such as

$10^{12} - 10^{10} = 10^{12}$. That equation is impossible if taken literally, because subtraction by definition gives you less than what you started with. What the equation really means is that at any given order of magnitude, the lower ones might as well be zero. If you start with trillions and subtract merely tens of billions, you're still left with trillions.

Former Senate minority leader Everett Dirksen made a name for himself by saying, "A billion here, a billion there; pretty soon you're talking real money." (Or at least people say he said it. Scholars have combed 12,500 pages of Dirksen speech notes without finding any such reference.) But billions don't become trillions without a lot of time and effort.

Just ask Bill Clinton. When the Clinton administration began paying off the national debt in 1999, the symbolism was enormous but the near-term effects were limited: In the twelve months that followed, during one of the rosiest periods in U.S. economic history, debt repayments amounted to $300 billion, still a small number in the context of $5.7 trillion worth of debt. (Note the silent juxtaposition of different orders of magnitude. We're conditioned to tune out numbers that follow a decimal point, but the ".7" part of $5.7 trillion is itself $700 billion.) The "Panetta Plan" called for elimination of all federal debt by 2015, but achieving that goal required that the initial pace of debt retirement be sustained over a 16-year period. In the event, surpluses lasted but another two years; by 2002 the federal budget was once again in the red.

If we delve deeper into the federal budget, we see orders of magnitude represented in the form of Pareto's Law. The government's revenues for fiscal 2001 totaled $1.99 trillion while expenditures came to $1.86 trillion, the net being a surplus of $127 billion. Here are the five big-ticket items on the expense side:

Category	Outlays
Social Security	$433 billion
Health (including Medicare)	$390 billion
National defense	$309 billion
Income security	$270 billion
Net interest on debt	$206 billion

Adding up these five categories produces a price tag of $1.6 trillion, over 85 percent of all expenditures, just the type of domination that Pareto's Law would predict. That's right: All told, every other area of spending—education, training, employment, and social services, veterans benefits and services, energy, natural resources and environment, commerce and housing credit, transportation, community and regional development, international affairs, general science, space and technology, agriculture, administration of justice, and general government expenditures—amounts to less than 15 percent of total federal outlays. Small wonder that economist Paul Krugman referred to the federal government as "a big pension fund that also happens to have an army."[7]

What do these figures mean? From a policy standpoint, they mean that if you want to shrink or restrain the growth in the budget, you begin by investigating the five big categories. While the buzzwords of the nineties economic boom were productivity and telecommunications, the shrinking of the defense budget during this same period, made possible by the end of the Cold War, allowed for steady and meaningful (but in some sense underappreciated) reductions in the expense side of the federal income statement. But in real terms (meaning adjusted for inflation), defense spending bottomed out in 1998 and nobody expects another major decline anytime soon. So looking ahead, the size of the federal budget will depend, more than anything, on the growth in outlays for Social Security, Medicare, and Medicaid.

Whatever the future has in store for the five biggest expense categories, we are left with two symmetric and not sufficiently appreciated points. The first is that the budget of the United States of America is not, as is believed in some cynical circles, dominated by wasteful, pork barrel spending programs. Yes, there is waste in our government, as with any government or any organization of that size, but the actual amount of waste is perhaps an order of magnitude lower than the popular perception. The symmetric point is the corollary that the vast majority of budget maneuvering and political posturing takes place within a relatively small subset of federal spending, where dollars often take a back seat to symbolism.

Foreign aid provides some vivid illustrations of symbolism at work, and we'll begin by turning the clock back to 1989. When President George H. W. Bush visited Poland that summer, he proposed an aid package of $119 million, saying it was all that America could afford. If that alibi sounds fishy to you, you have good company in the person of ex–Federal Reserve chairman Paul Volcker. Upon hearing Bush's pronouncement, Volcker was quick to recall his Colorado vacation of earlier that year and point out that the aid package amounted to about 30 luxury homes in Aspen. However, even if the United States could have "afforded" to give Poland $10 billion, such a gift might not have been good policy. Besides, the symbolism of the aid package meant as much as the amount, because there weren't a lot of precedents for the United States giving money to a Communist state. The world was changing, and when you're starting from a zero base, $119 million is a big change.

But that's only a single anecdote in a much larger story on foreign aid. For the record, as the twenty-first century began, the

United States was earmarking on the order of $9 billion per year for total foreign aid, whether bilateral or multilateral. Is this a big number? Yes and no. In absolute terms the U.S. trailed only Japan, which was giving out $15 billion a year despite being gripped in economic stagnation for more than a decade. However, when you assess the generosity of the U.S. relative to the size of its economy, you see that we are comparatively stingy; that $9 billion figure is merely 0.10 percent of GDP. By comparison, Denmark gives 1 percent of its GDP to foreign aid. That's right. Denmark's foreign aid, as a percentage of GDP, is ten times that of the U.S.

Not that the United States should necessarily try to keep up with Denmark and dole out something on the order of $90 billion in foreign aid each year. There are serious questions about the effectiveness of foreign aid in helping its intended recipients. However, by the standard of other industrialized countries, we could obviously contribute more than we do. If you ranked all countries in terms of the foreign aid they provide relative to GDP, the U.S. wouldn't rank in the top 25.

Fortunately for the U.S., a legacy of the postwar Marshall Plan is that in some circles the U.S. is still thought of as being remarkably generous. Admittedly, this perception may be stronger within America's borders than outside it, but it seems safe to say that there is no political imperative for the U.S. to beat out, say, Saudi Arabia so as to move into 24th place on the top percentage givers list. Here again, though, we face a perspective issue. With Saudi Arabia's foreign aid at 0.14 percent of GDP, it looks as though the U.S. could easily make up the difference, but the old "percent of a percent" game intervenes. Any way you slice it, going from 0.10 percent to 0.14 percent requires an increase of 40 percent.

The plain truth surrounding foreign aid and many other government expenditures is that precisely because our ears aren't tuned to differences in order of magnitude, our leadership can make symbolic statements in the millions—enough to get people's attention, but effectively a zero in the arithmetic of orders of magnitude. The principle of proximity strikes again.

The Trouble with Small Numbers

We all know that the chance of getting killed in commercial air travel is small. How small? To answer that we need to figure out how to measure mortality risk in a way that provides useful perspective. That is to say, we need to measure mortality risk in a way that's easy to understand and gauge, and accurately characterizes the danger of flying.

One common measure is the number of fatal accidents per 100,000 flight hours. Between 1982 and 2001, according to figures compiled by the National Transportation Safety Board, scheduled U.S. airline service averaged 0.3 fatal accidents per 100,000 flight hours. Yet MIT professor Arnold Barnett and management consultant Alexander Wang note that there are two problems with that measure: the numerator and the denominator.[8] The numerator, fatal accidents, "does not distinguish between an accident that kills one passenger among 300 and an accident that kills everyone aboard. The term gives no credit to safety improvements (fire-retardant materials, for example) that reduce fatalities but do not prevent them." The trouble with the denominator, flight hours, is that it gives excessive weight to longer flights. Mortality risk is not proportional to the length of a flight, since most accidents occur during the takeoff and approach-and-landing phases.

Barnett and Wang suggest that a more useful measure of mortality risk is obtained by looking at flight safety from the perspective of a traveler. And so they ask the following question: "If a passenger chooses a (nonstop) flight completely at random, what is the probability that he or she will be killed during the flight?" On their calculations, using data on the scheduled domestic jet flights of U.S. carriers between 1987 and 1996, the answer is 1 in 7 million, or 0.00000014. But is a probability of 0.00000014 big? Barnett and Wang recognize that such a small number is hard to make sense of and nicely convert it into a statistic that is more readily grasped: A passenger who randomly flew on a U.S. domestic jet every day would go, on average, approximately 19,000 years before dying in a fatal accident.

In our experience, small numbers (such as 0.00000014) are even more difficult to get a handle on than big numbers. We can think of four reasons for this and being aware of them may help you deal with small numbers.

First, small numbers are far less common than big numbers. If you look through a newspaper, you will come across millions, billions, and sometimes trillions, but rarely will you encounter a positive number less than 0.001.

Second, small numbers imply division, while big numbers imply multiplication. 7,000,000 suggests $7 \times 1,000,000$ while its reciprocal, 0.00000014, suggests $^{14}/_{100,000,000}$. As any third grader can tell you, division is less intuitive than multiplication.

Third, outside of scientific copy, there is no well-known convention for grouping the digits of a small number in a way that makes its magnitude easy to identify. 7000000 is 7,000,000, but 0.00000014 is just 0.00000014. You might see 0.000 000 14 in scientific copy, but in any other written material you'd wonder if a typographical error had inserted two spaces or dropped a couple of digits.

Fourth, most people not only lack a notation for dealing with small numbers, they also lack a vocabulary. If you show a man the number 3,500,000 and ask what it is, he will say, with little if any hesitation, "three and a half million," "three point five million," or "three million five hundred thousand." But if you show him 0.00000029, he will probably respond, "point oh oh oh oh oh oh two nine," slowly, as his index finger passes over each digit. And, frankly, what should we expect him to say? "Twenty-nine hundred millionths?" The meaning of which would be unclear, anyway. Is that "2900 millionths" or "29 hundred-millionths"? "Two point nine times ten to the minus seventh"? Not likely.

How can we expect people to grasp the magnitude of 0.00000029 when they don't even have a term for it? You may not realize just how empty "point oh oh oh oh oh oh two nine" is. But just imagine a politician describing the defense budget as "three six nine oh oh oh oh oh oh oh oh dollars."

How Small Things Become Big

In seeking labels of "small," "medium," and "large" for the numbers we investigate, we should be aware that numbers have been known to change their labels over time. If you remember the old public service advertisement, "Every litter bit hurts," you are reminded that little things do indeed add up.

To echo some of our remarks from chapter 2, the ability of small things to accumulate is in no way a repudiation of Pareto's Law. If you were responsible for monitoring our environment as a whole, you wouldn't start with littering; you'd start by trying to make sure that nothing like the Exxon *Valdez* ever happened again. A single slip-up of that magnitude creates more damage than a million litterers working around the clock, but that doesn't argue not to crack down on littering. Similarly, if you're the class agent trying to drum up contributions from the Harvard College class of 1977, Pareto's Law makes you especially interested in envelopes from your classmates Bill Gates and Steve Ballmer, but the smaller checks are meaningful symbolically and do add up, even for Harvard. And whereas writers are paid to fill up pages with words, we note that our publisher could have made this book 20 percent thicker with a different choice of font size, line spacing, or even paper size.

Whatever you make of little things adding up, they can add up much more quickly if we can form *combinations* of small quantities. For example, when you consider that a deck of playing cards numbers only 52, you'd think the game of bridge would get very stale, very quickly. Consider that there are 3,500 duplicate bridge clubs sanctioned by the American Contract Bridge League, or ACBL. If each club holds even as few as two duplicate games a week, that's a total of $104 \times 3,500 = 364,000$ games per year. An average duplicate game consists of 24 deals, so the total number of deals in club games alone—never mind tournaments or undocumented informal play—comes in at 8,736,000 per year. But that's still a puny number when measured against the number of ways in which 52 cards can be arranged among four players, a figure that comes in at a staggering 54 octillion. As much as we hate the phrase "almost infinite," the number of bridge deals is without a doubt functionally infinite, because even a lifetime of playing will never get you close to saturation. Even if the demographic profile of the game weren't declining (a polite way of saying that players are dying off), the vast majority of bridge deals would never, ever get played, a notion that somehow remains counterintuitive despite its probabilistic certainty.

The world of games is filled with examples of combinations and huge numbers. If you take six standard 2×3 Lego bricks, they can be put together in over 100 million different ways—

some, of course, more stable than others. More astounding is the number of possible positions for a Rubik's cube, which comes in at an inconceivable 4.3×10^{19}, a figure at least 40 times bigger than the number of seconds in the history of the universe.

Sensitive Numbers

A different way for small numbers to yield big numbers is if one or more numbers in a calculation are sensitive. We admit that the phrase "sensitive number" looks peculiar, because numbers don't have feelings. When performing calculations involving many different numbers, the sensitive numbers are the ones for which small changes have the capacity to greatly affect the entire calculation. Sensitive numbers, like sensitive people, must be handled with all the delicacy you can muster.

To illustrate that numbers are not equally sensitive, recall the Rubik's cube figure that we just saw: 4.3×10^{19}. Three numbers are involved: 4.3, 10, and 19. Let's consider the impact of increasing each of them by 10 percent. Raising the 4.3 by 10 percent would simply increase the original number by 10 percent, to 4.73×10^{19}. Alternatively, raising the 10 by 10 percent would give us 4.3×11^{19}, which is 512 percent larger than the number we started with. And raising the 19 by 10 percent would change the original number to $4.3 \times 10^{20.9}$, an increase of 7,843 percent. Clearly, exponents and numbers related to them can be very sensitive.

Denominators are also known for their sensitivity, especially when they are 1) small, and 2) rounded off. One plentiful source of rounded-off numbers is *Pocket World in Figures*, a nifty little reference book produced each year by *The Economist*. In the 2002 edition, we learn on page 66 that Somalia and Tajikistan are tied for bottom in the category of per capita car ownership. Both countries have just 0.1 cars for every thousand people. (By comparison, the U.S. ranks ninth with 486 cars per thousand people; Lebanon, with 732 per thousand, is at the top of the list, a ranking that frankly puzzles us.) However, because the figures for each nation are rounded to the nearest tenth, that 0.1 figure could be anything from 0.05 to just under 0.15. Even knowing that there are 6.2 million people in Tajikistan (from the country summary on page 237), the actual number of cars could be anywhere from 310,000 to 929,999, because we'd get the same rounded-off percentage ownership (0.1 percent) for any number in that range.

Before you scream "Who cares about the number of cars in Tajikistan?" we should reveal that a similar type of "multiplier" effect is found in that favorite number of stock market investors, the price/earnings ratio. The price/earnings ratio, or P/E, is just what its name implies. It is calculated by dividing a company's stock price by its earnings per share. A $30 stock with earnings per share of $1.50 sports a P/E of $^{30}/_{1.5} = 20$. The P/E construction is necessary because the absolute price of a stock—whether it be $30, $90, or even $5—isn't what determines relative value in the marketplace. If a company with a $5 share price is earning only $0.10 per share, the stock is arguably more expensive than the $30 stock above, because the P/E for the $5 stock is $^{5}/_{0.1} = 50$.

Already, though, we've glossed over a couple of problems. Does the earnings per share figure refer to last year's earnings (the most recent hard numbers), this year's anticipated earnings (trying to be more current), or next year's estimated earnings (trying to look ahead)? Suppose that the $1.50 per share figure above was a current-year estimate, up nicely from last year's $1.20 per share but a far cry from the $1.80 per share anticipated a year from now. Your P/E could therefore be 25, 20, or 16.7 depending on which base you choose—and all three choices are in widespread use. In particular, if a company is growing quickly, you can create a much lower P/E by using projected earnings, but you also add a risk that those projected earnings won't materialize.

But an even bigger problem arises when we try to perform the essential task of comparing the P/E of an individual stock with that of the market. What's the market P/E? It depends on whom you ask. The price of an index such as the S&P 500 is known at every moment, but defining earnings for an index of stocks is considerably more complicated than the already complicated task of defining earnings for an individual company, in part because some firms have negative earnings and there are differing views as to how such losses should be counted. The bottom line is that *the* market P/E doesn't really exist as a single, well-defined number. At any given moment the market multiple might be 15, 20, 30, or even 50, all according to differences in methodology that are almost always invisible to the poor schnooks (that's us) who encounter the number in the financial pages. So, back to our $30 stock trading at a P/E of 20. Is it cheap? Expensive? Somewhere in between? Suddenly the most important valuation question you can ever ask has become little more than a guess, all because of differing methodologies and the sensitivity of the denominator.

Survivorship Bias

It's time for a quick breather. The importance of perspective isn't restricted to numbers. Parents have been known to remind their whining kids that melted ice cream doesn't constitute a genuine tragedy. And sometimes adults also need to be reminded to look beyond their immediate circumstances. In his book *Fooled by Randomness*, hedge fund manager Nassim Nicholas Taleb cites a lawyer living in a pricey co-op on Park Avenue. Here is a person with a lifelong record of overachievement. He outperformed virtually all of his high school classmates, hence his acceptance to Harvard. He outperformed 90 percent of his Harvard classmates, hence his acceptance to Yale Law School. He then outperformed 60 percent of his law school classmates, hence his position with a major New York firm. Alas, despite all of these academic accomplishments, he considers himself a failure because he remains at the bottom of the heap when compared to the other residents of his Park Avenue co-op.

We needn't shed any tears for this fictitious fellow, but he exemplifies what academics call "survivorship bias." By the time he reached Park Avenue, he had already "outlived" numerous layers of competition, at which point he measured himself only against those who had done the same, with psychologically disastrous consequences.

One real-life story along these lines was experienced by famed Mexican tennis player Joaquín Loyo-Mayo. Actually, he isn't that famous, which is part of the point. While attending USC, Loyo-Mayo (pronounced Loy-oh My-oh) won the U.S. intercollegiate title in 1969, one of the two players sandwiched between a couple of intercollegiate champions named Stan Smith and Jimmy Connors. But when Loyo-Mayo hit the pro tour, he also hit the wall. At that level of the game his diminutive physique became a distinct handicap and he basically couldn't beat anybody. So what happened? He went home to Mexico, and, by now accustomed to losing whenever he stepped on the court, found that he couldn't beat players who two years before might have asked for his autograph.

The point is that comparing ourselves to others is always risky business, whatever the nature of the sport or competition. Dave Pelz has made millions from the game of golf as the author of *Dave Pelz's Putting Bible* and *Dave Pelz's Short Game Bible*, but he never made it on the pro tour. Long before he committed himself to a career, he realized that his chances as a golf pro were nonexistent; after all, while on the Indiana University golf team, he lost over and over to a fat kid from Ohio State. That the "fat kid" was named Jack Nicklaus didn't seem relevant at the time.

Most of us have probably had similar experiences, albeit on a smaller portion of the world's stage. What isn't easy to see is that survivorship bias has a numerical formulation in the world of mutual fund performance. Suppose you wanted to get a feel for the performance of funds in specific industries, such as technology or health care. You might look at ten or so funds in each category. But those are the *survivors*; the mutual fund tables in the newspaper don't show you all the funds that people invested in five or ten years ago that subsequently closed their doors because of poor performance. With the Park Avenue example fresh in our minds, we can see that by looking at only the surviving funds, we get a rose-colored picture of overall mutual fund performance. Rest assured that this survivorship bias has not escaped scrutiny from the academic community. Studies have shown that within such volatile (and failure-prone) categories as hedge funds and commodity funds, the inclusion of failed funds in performance calculations reduces the group's apparent returns by two or three percentage points a year.

A different type of bias arises when we seek out low P/E stocks. Sure, it's great when we come across a solid company that just happens to be trading at a sharp discount to the market; in the best of worlds, earnings will improve and so will the P/E, giving investors a nice "double play" for their efforts. But the universe of low P/E stocks by definition includes companies whose future is being questioned by the market, either because of a downturn in their basic business or because they carry too much debt. In other words, low P/E stocks include both future survivors and future casualties, and so investment performance in that segment of the market depends heavily on being able to separate the wheat from the chaff.

To round this story out, an unexpected link between investing and surviving emerged during the *Barron's* investment roundtables of the mid-eighties. When pressed for a few favorite stocks, fund manager Mario Gabelli came up with some restructuring candidates among stodgy, low-growth companies whose main appeal was that the chairman or primary shareholder was of such advanced age that a shift in ownership was inevitable. The sooner an "exit strategy" is carried out, the better it is for everyone—except perhaps the chairman. Okay, let's shed our fears of mortality and hop right into the question. Assuming that a man's average life span is 74 years, what is the life expectancy of a chairman of the board (or any man at all) who is 65 years old? The answer, according to tables used by none other than the Internal Revenue Service, is 15 years. Note that life expectancies are given in terms of remaining years, reminding us that the 74-year expectancy applies only at birth; it lengthens with time, as survivorship bias would suggest. Even at age 75, male life expectancy is 9.6 years. At 90, the IRS still gives a man 4.2 more years, and at 100 he has 2.1 years. A man's life expectancy doesn't dip below one year until he's 107 years old.

So much for Gabelli's personal form of ambulance chasing. As a postscript, when Earl Scheib, founder of the eponymous auto painting chain, finally got around to dying in February 1992, Scheib stock soared 50 percent on the news. Alas, Gabelli's December 1985 touting of the company and its 77-year-old chairman was without reward; Scheib had taken over six years to die, and his stock had declined 70 percent during the wait.

The Tactics of Big Numbers

To close this chapter, we'll return to our original point that not everyone is facile with orders of magnitude. This simple observation has important implications when it comes to the tactics of using numbers.

We are reminded of a longtime restaurateur in the Wall Street area who saw the benefits of serving breakfast to some of the most powerful dealmakers in the financial community. What better way to kick off a day of multimillion-dollar deals than with an $8 glass of orange juice? The clientele eventually balked, noting that orange juice should be priced relative to orange juice, not arbitrage profits, but even the compromise price was immensely profitable.

Sheridan Whiteside of *The Man Who Came to Dinner* was modeled by playwrights George S. Kaufman and Moss Hart on bloated egocentric critic Alexander Woollcott, but Whiteside also showed a quantitative flair. His opening foray was to threaten his host Mr. Stanley with a $150,000 lawsuit for some nonexistent injuries suffered in a fall. Then, when Stanley dared complain about the $784 in phone charges rung up by his unwanted guest, Whiteside's counteroffer was to subtract $784 from the $150,000.

Both the Wall Street restaurateur and Sheridan Whiteside attempted to make big numbers—an $8 glass of OJ, a $784 phone bill—appear small by juxtaposing them against much larger numbers. By using bogus reference points, they tried to have numbers slip by unnoticed.

Had they wanted to call attention to their numbers, a different set of tactics would have been required. Specific numbers are more memorable than round numbers. A few pages back, we noted that six standard Lego bricks can be put together in over 100 million ways, an order of magnitude much higher than one might expect. Now suppose you worked in Lego's marketing department and wanted your salespeople to remember that fact. You wouldn't give them the exact number, 102,981,500, which you will have forgotten by the time you turn this page. But providing the rounded-off figure of 100 million would be risky for different reasons, because in a world where such figures get tossed around ad nauseam, they automatically lose their individuality. Was that one hundred million, a million, a couple hundred thousand? The hell with it, it was just a big number, right? In this instance, citing the number 103 million is probably the best bet for retention.

When it comes to actual decisions by game companies, perhaps the best known is that of the Ideal Toy Company, which released the first batch of Rubik's cubes with the stunning revelation that there were more than 3 billion combinations. Douglas Hofstadter, then writing for *Scientific American,* was quick to point out how ridiculously low that number was. Recall that the number of combinations turned out to be 4.3×10^{19}, a number that not only leaves 3 billion (3×10^9) in the dust, but is also bigger than 3 billion squared. Hofstadter's memorable phrase for Ideal's approach was "a pathetic and euphemistic understatement," but clearly Ideal was using the principle of proximity 20 years before that term saw the light of day. Unlike 4.3×10^{19}, the number 3 billion was well within the range of familiar numbers, and already so huge that there was no extra mileage to be had from revealing the precise figure. And with 4.3×10^{19} only describable by use of an exponent, its fate was sealed. Oh, well. At least they didn't say "almost infinite."

You don't have to delve into such gargantuan numbers to see the "less is more" doctrine at work. When your mother said, "I've told you a million times not to do that," what was your reaction? Did you like it? Did you say, "Good point, Mom, I'll get right on it"? Somehow we don't think so. Even a first-year child psychologist understands that if you repeat the same message more than three times without effect, it's time for a new message. Perhaps Mom should have settled for several orders of magnitude lower. "I've told you *twice* not to do that" is in many ways a more powerful statement, starting with the fact that it is believable. In the immortal words of Alfred E. Neuman: "Understatement is a zillion times more effective than exaggeration."

Throwing a Curve

The world isn't flat. When we try to make it flat, as on a map, the result is distortion. Greenland is not actually bigger than Africa, no matter what our old world map seemed to suggest. Maps inevitably alter reality by trying to flatten the natural curvature of the earth. A similar type of distortion exists in the world of numbers. All too often we assume that progressions of numbers are linear instead of curved.

The tendency is understandable, since numbers are so much easier to deal with when they follow a nice, straight-line pattern. If you earn five dollars an hour and work for two hours, you've earned ten dollars. Three hours means fifteen dollars, four hours means twenty dollars, and so on. Excluding the wrinkle of overtime pay, your total earnings move in lockstep with the number of hours worked. Predicting future income streams is a breeze.

Having extolled the virtues of taking the easy way out, we'd love to approach other calculations the same way. But look at the trouble we'd run into. An average American boy attains half of his eventual height after 35 months of growth (including gestation), at which point he weighs 28½ lbs. If growth were linear, he would end up 5'10" and 57 lbs—at age five. Growth curves capture the reality that relationships among height, weight, and age are anything but orderly. A newborn will typically grow something like 10 inches and 14 lbs in year one. By contrast, a two year old will normally put on about 3½ inches and 4 lbs in a year, while a ten year old will usually gain around 2 inches and 10 lbs.

The habit of looking for the easy way out only works when you understand something well enough to know which shortcuts are worth taking. That's where we'd like to go in this chapter, but first we have to train our linear brains to deal with the world's built-in curvature.

Modeling Through Hidden Curves

In the course of writing this book, we came across a curious newspaper headline: "Waiting times to be cut in half." According to the article that followed, a hospital wanted people to know that it was doing something about people waiting in line to receive emergency care. Their antidote, it turned out, was to double the staffing during certain shifts. Their conclusion was that people would wait only half as long. Good news indeed, and worth spreading.

But who was doing the PR here? Why would the hospital let its efforts be summarized by a headline that, in addition to being wimpy, was flat-out wrong? The fallacy of the hospital's waiting time claim isn't hard to sort out. Consider a simple model that could apply anywhere people queue up for anything: one line, one staff member, one new customer every five minutes, and an average time per customer of ten minutes. What would happen? In the next hour, twelve new patients would have come in, but only six would have been attended to. The number of people in line would be six, and that number would be growing without bound.

Enter a second staff member. Now, on average, you'd have complete equilibrium. Every ten minutes two customers would come in, and there would be two staffers to process their requests. Yes, there might be unusually busy times, but for the most part, there would be no lines to speak of. Instead of merely cutting the waiting times in half, a doubling within this simple model would eliminate the waiting times altogether.

What we found interesting about the waiting time blunder is that it defied a nasty stereotype. When companies misuse numbers, the automatic assumption is that they are acting on their own behalf and that any "mistakes" are actually a form of corporate malfeasance. In this case, however, the abuse of numbers was demonstrably to the *disadvantage* of the company involved. The benefit to consumers was greater, not less, than the hospital claimed, all because whoever crafted the hospital's response was living in a linear world.

Mistaken linearity is at the root of a classic conundrum of average speeds, which goes something like this: You drive from point A to point B going 20 miles per hour. How fast do you have to go on the way back to average 40 miles per hour for the whole trip? The tempting answer is 60 miles per hour, but in your heart of hearts you know that can't be right. Again, the way to sort things out is with a simple model. If A and B are precisely 20 miles apart, the round trip is 40 miles, so if you want to average 40 miles per hour, you had better complete the round trip in precisely one hour. Unfortunately, if you've piddled along at 20 miles per hour for the first half, you've used up your entire hour. That's right. No return speed, not 60, 120, or even 240 miles per hour, could possibly produce an average speed of 40 miles per hour.

This paradox of average speeds should be familiar to anyone who's been on a highway trying to make up for a late start. If your destination is a couple of hours away, then driving a little faster can make up the ten minutes or so you lost at the beginning. But if you want to make up that time over a short span, forget it. Driving 30 miles at 60 miles per hour will take you 30 minutes. If you want to do it in 20 minutes instead, you have to go 90 miles per hour.

There *is* linearity within the average-speed paradox: If you drive for a while at 20 mph and then drive that same amount of *time* at 60 mph, your average speed will indeed be 40 mph. But real-world situations are more likely to be defined by a common distance, as on a round-trip bicycle ride. If you ride your bike 5 miles uphill, you'll be happier on the downhill return, but when all is said and done you'll have spent a clear majority of your time going uphill, which is why an increase in your uphill speed will do more to improve your total time than an increase in downhill speed.

Scales and indexes are usually designed to make our lives easier, but they can introduce some surprise curvature along the way. For example, Charles Goren made a name for himself in the 1950s by introducing the concept of "point count" for a bridge hand. In Goren's system, an ace is worth four points, a king is worth three, a queen two, and a jack but a single point. The point-count system remains inextricably woven into the fabric of the game, but it has many flaws. In particular, jacks are overrated by Goren's approach, while aces are vastly underrated. That and other shortcomings prompted modern bidding expert Marty Bergen to write a book called *Points Schmoints!*

The Richter scale measures the intensity of earthquakes on a scale from 1 to 10. By now the scale is fairly institutionalized, in that when a major earthquake strikes, you can be certain that the various news anchors will recite the associated Richter reading. Unfortunately, the Richter scale number is usually delivered more as an obligation than as an insight. Most viewers have no way of knowing that the scale is based on the *logarithms* of seismological readings. What that means is that an extra point on the scale acts like a zero at the end of a number, increasing the strength of the earthquake by a factor of ten. Only with this information can we achieve any sort of perspective. For example, the calamitous earthquake that hit Uttarkashi, India, in 1991 was assigned a 6.1 by the Indian Meteorological Department, while the U.S. Geological Survey assigned a 7.1. In other words, there was a tenfold difference between these seemingly close measurements.[1] But even the larger figure of 7.1 is only 1/100th as big as the 9.1 that the USGS gives for the magnitude of the quake that struck the Andreanof Islands of Alaska in 1957.[2] In essence, the simplicity of the Richter scale is precisely what makes its output so misleading.

The pH scale operates in much the same way. If the pH readings of two substances are a point apart, the one with the lower number is ten times as acidic as the other. The pH scale rarely makes headlines, but we can't help but wonder how many aquarium owners have said, "Close enough," when monitoring pH levels in their tanks, only to have their fish die in overly acidic water.

At the Margin

When drivers from *Road & Track* magazine tested Chrysler's popular PT Cruiser, they needed 131 feet to stop the car from a speed of 60 mph. However, when the same test drivers slammed on the brakes at 80 mph, the stopping distance was 232 feet.[3] That an extra 20 mph added so much to the Cruiser's skid marks may surprise you, but it wouldn't have shocked Sir Isaac Newton. The classical mechanics he developed predict that minimum stopping distance increases with the square of velocity. Therefore, Newton would have estimated that a car requiring 131 feet to stop from 60 mph would take 232.9 feet to stop from 80 mph $\left[\frac{80^2}{60^2} \times 131 = 232.9 \right]$. Pretty good guess for a guy who predated the motorcar by two centuries. Then again, Newton did understand a thing or two about nonlinear relationships.

Using braking as an illustration, let's give nonlinearity a closer look. As a simple model, suppose that the minimum stopping distance of your car, measured in inches, equals the square of its speed in mph. At 50 mph, minimum stopping distance is 2,500 inches; at 60 mph, it's 3,600 inches, and so on. It's unlikely your car has such lousy brakes, but remember, this is just a model.

Say you're driving at the grand speed of 1 mph. It would take all of an inch to screech to a halt. At 2 mph you would need 4 inches to stop, or an additional 3 inches. From 3 mph, stopping would require 9 inches, a further increase of 5 inches. This pattern is what makes stopping distance nonlinear: Each additional unit of speed increases stopping distance by more than the previous unit did. The first mph added one inch to stopping distance; the second mph added 3 inches; the third mph, 5 inches.

What we're talking about here is *marginal* change. When a relationship is linear, every marginal change has the same effect. At five bucks an hour, an extra hour of work yields an additional $5 in pay, regardless of whether it's the first hour of work or the thirtieth. When a relationship is nonlinear, marginal change is not constant. Accelerating from 30 to 31 mph increases stopping distance by 61 inches in our model, while speeding up from 60 to 61 mph adds 121 inches.

Paying careful attention to marginal change is critical to understanding and dealing with a numerical world. Many of the questions we pose in life require an assessment of marginal change. What will happen to a company's earnings if its sales increase by $100 million? By how much will crime rise if a city cuts its police force by 3 percent? Should I have an another slice of pizza?

Sloppy reasoning about marginal change will often lead to poor answers to such questions. A standard mistake is to assume that marginal change is constant when it's not. If you're eating pizza, it's obvious that a fifth slice wouldn't add to your satisfaction the way the second slice did, but people often fail to apply that lesson in other circumstances. For example, some house hunters don't recognize that the marginal cost of heating or cooling additional square footage declines as homes increase in size. Another classic error is to assume that marginal change is constant in percentage terms. If a company earns $50 million in profits from $1 billion in revenues, then an additional 10 percent in revenues would add 10 percent to the bottom line, right? Probably not. Assume the company has $500 million in fixed costs—that is, costs that are unrelated to its volume of business—and another 45 cents in costs connected with each additional $1 in revenues. Then an extra $100 million in revenues would entail $45 million in costs and generate $55 million in profits, more than doubling the company's earnings. Now you understand why the earnings of companies that invest heavily in factories and other equipment—the semiconductor business comes to mind—are highly sensitive to changes in revenues.

When thinking about marginal change, you should always ask yourself whether marginal change is increasing or decreasing. Sticking with the commercial world for a moment, imagine that a small printing company hires a salesman to drum up local business. The salesman earns his keep: the revenues he brings in exceed his compensation and the costs of doing the new work. How would a second salesman fare? Likely, his marginal contribution to the company's profits would be less than the first salesman's. Even if the two don't get in each other's way, the first salesman has presumably targeted the best prospects or territory, leaving the second one to pursue less attractive opportunities.

At some point, the marginal benefit of additional sales efforts declines to zero and then turns negative. Identifying the point of zero marginal benefit is key; in principle, it signifies the optimal level of sales work. That's Business 101 stuff, but the idea that shifts in marginal change can reveal these kinds of optimum levels applies to almost every facet of life. For instance, if you avoided business courses and took Psychology 101 instead, you would have encountered the Yerkes-Dodson curve, which is generally presented as showing the relationship between emotional arousal (anxiety) and performance. The curve is shaped like an inverted U, implying that increased arousal improves performance—up to a point. After that optimum, the marginal effect of heightened arousal is negative, and increasingly so.

If you've followed this discussion of marginal change, you've done more than you think. Not only are you on your way to understanding the ups and downs of the world, as a bonus, you've also grasped the essential concepts of the first semester of calculus. No fooling. Calculus is widely considered "advanced" math, but the initial ideas are precisely the ones we've just covered. Marginal change; change in marginal change; and optimum values. That's it. Really.

Square Curves

Lines all have the same shape: straight. Curves, however, can take on an infinite number of shapes, which is a good part of what makes nonlinearity so challenging. Yet despite their endless variety, most curves do not follow random patterns. Some patterns show up so often that it's worth exploring their particular features.

Braking distances aren't the only thing about your car that's not linear. To make your car go 75 mph instead of 50 mph, your engine doesn't simply push 50 percent harder, but about twice as hard. That's because air resistance, the dominant force opposing the motion of a car at highway speeds, increases with the square of velocity. Air resistance at 75 mph is 2.25 times air resistance at

50 mph $\left[\frac{75^2}{50^2} = 2.25\right]$. This nonlinear link between velocity and aerodynamic drag is the reason speeding reduces fuel efficiency, which in turn is why many conservationists opposed repeal of the national 55 mph speed limit.

Note that the relationship between speed and air resistance, much like that of speed and braking distance, involves squaring one quantity or another. These so-called quadratic or second order relationships are important because they're all over the place. Some are easy to spot: When you double the side of a square, you don't double its area; you quadruple it. That's obvious enough to anyone who has ever looked down at kitchen or bathroom tiles, a small portion of which might look like this:

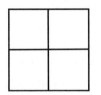

Other times we have to do a bit of work to see the quadratic connection. When you double the diameter of a circle, the area is quadrupled, just as with a square, but that inference isn't quite as clear to the eye:

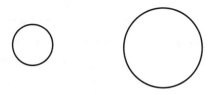

To reach the desired conclusion, you can use the formula for the area of a circle, $A = \pi r^2$, which exhibits the squaring directly, or you can nest those same circles in squares and note that each circle/square diagram is in proportion.

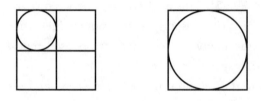

Measurements for television and computer screens are usually given as one number: A 15-inch monitor measures 15" along the diagonal. But, as we've seen before, there's quadratic curvature lurking within this linear index. If you upgraded from a 15-inch to a 20-inch monitor, that's a 33.3 percent increase in the diagonal measurement. Assuming the old and new screens are the same shape, the height and width have also increased by a third. Thus the overall size of the screen has gone up by a factor of 1.333 × 1.333 = 1.777, or by 78 percent.

Other quadratic relationships are more obscure and therefore even easier to overlook. For example, it would be natural to assume that the strength of a combat force is a linear function of its size. But military planners have long assumed that in many combat situations, the strength of a force is related to the square of its size. This is a feature of a model known in military circles as Lanchester's Square Law.

Pretend the Good Guys are fighting the Bad Guys and each side has 100 tanks. The Good Guys and the Bad Guys have equally effective tanks, and every time a tank fires at another tank, there's a 10 percent chance that the fired-upon tank gets knocked out of commission. Thus, after one round of firing by all the tanks, we can expect that both sides will have suffered a 10 percent attrition rate.

New battle. This time the Bad Guys have 200 tanks while the Good Guys still have 100. As a result, the Good Guy tanks face twice as much fire as they did in the previous battle, roughly doubling their attrition rate to 20 percent. On the other hand, each Bad Guy tank faces only half as much fire as it did when the forces were equivalent in size, since there is now only a half a Good Guy tank per Bad Guy tank. The Bad Guy attrition rate falls to 5 percent, one-fourth of the Good Guy loss rate. Thus, a doubling of force size has a fourfold effect on relative attrition rates. The reason for this nonlinear relationship is that there are two effects when the Bad Guys increase their force size: The Good Guy attrition rate goes up and the Bad Guy attrition rate goes down.

Third battle. The Bad Guys again have a 200 to 100 edge in force size. However, the Good Guys, in an effort to overcome their numerical disadvantage, have deployed tanks that are four times as effective as conventional tanks. When a Good Guy tank fires, it now has a 40 percent chance of incapacitating its target. That quadruples the Bad Guy attrition rate to 20 percent. But it doesn't change the rate of Good Guy losses, which remains at 20 percent. The forces are now evenly matched. A doubling in force size has the same effect as a quadrupling of combat effectiveness.

The implications for military strategy are obvious. During the Cold War, many Americans felt reassured by the technical superiority of NATO's conventional forces in Central Europe, figuring it would counterbalance the Warsaw Pact's substantial numerical advantage. Pentagon planners, aware of Lanchester's Square Law, did not rest so easily.

If quantities can increase in quadratic fashion, they can also decrease quadratically. We don't have to look to the grand stage of the Cold War to find examples; there are plenty of illustrations right in your own home. If you grill a piece of meat or fish three inches from the heat source, the food will cook in about a quarter the time it would if you grilled it at a distance of six inches. The underlying physical principle is that radiant energy is related to the square of the distance from its source. But we're less interested in the physics than the practical effect; when you move something twice as far away from a grill, a broiler, a radiant space heater, or a lightbulb, the heat intensity drops by three-quarters.

Exponential Growth

Exponential relationships are arguably even more common than quadratic ones. We've all heard what a marvelous thing compound interest is. It's time to understand how and why.

Consider two mutual fund managers whose performance you wish to analyze. The first posted an average compound annual return of 11 percent over the past 15 years. The second did slightly better, with a compound annual return of 13 percent during the same period. At first glance, these figures appear quite close. But are they? If you had invested $10,000 in the first fund at the beginning of the period, you'd have had $47,800 after 15 years. $(1.11^{15} = 4.78$: We'll assume the investment was made in a retirement account so we don't have to worry about taxes.) With the second fund, your account would have finished at $62,500, around 30 percent higher. What this disparity means is that the numbers 11 and 13 aren't really close together after all, because they represent a compounding effect that is exponential, not linear. Quantities are being multiplied, not added. The scale of percentage returns operates a bit like the Richter scale, in that every extra point represents a significant change. Albert Einstein is said to have called compound interest "the greatest mathematical discovery of our time,"[4] and if Einstein can marvel at its power, so can we.

You could argue that the hypothetical numbers 11 and 13 were selected to make for a wide disparity. A choice of, say, 5 and 7 wouldn't have been as dramatic. True enough, but that puts us in a position to understand a common accounting gimmick that surfaced along with the rest of the corporate scandals of 2002. When stocks were going up all the time, few people objected when corporations used double-digit numbers in projecting the expected return on their pension investments. Many companies even outperformed this actuarial rate. But there was certainly an incentive to keep the actuarial rate high, because it theoretically enabled a company to claim that future pension benefits were adequately provided for when they weren't. A more conservative assumption about pension returns—say, 8 percent—would have better reflected the reality of the turning financial markets. But the difference between 8 and 11 doesn't look too meaningful unless you're attuned to the nonlinearity of the return scale.

Any number that gets compounded or otherwise repeated is by definition sensitive, provided you're talking about long-term intervals. For example, if the Congressional Budget Office estimates that the U.S. economy will grow at a 3.2 percent rate for 10 years rather than, say, 2.8 percent, that extra growth could easily amount to $500 billion in additional tax revenues over that 10-year period. Again, though, to the untrained eye and much of the American electorate, the difference between 2.8 and 3.2 does not leap off the page.

Evar Nering, professor emeritus at Arizona State University, points out an interesting conundrum concerning fuel supplies.[5] He starts by supposing that our supply of oil is 100 years, meaning that the world's oil would last 100 years if it were consumed at its current rate. If the rate of consumption increased by 5 percent each year, how long would the supply last? The answer, at first glance surprising, is that it would last only 36 years.

But the story is only beginning. Suppose you underestimated the supply rather badly, say, by a factor of 10. At the same annual 5 percent growth rate in use, how long would a 1,000-year supply last? Again, not as long as you'd think: just 79 years. Even an original 10,000-year supply would be reduced to 125 years under the simple condition of 5 percent growth. The point is that the growth rate number is far more sensitive than the supply number.

Nering's conclusion is that you can't solve the world's energy problems through the supply side alone. Of course, another interpretation is that the 5 percent growth figure is actually a huge number. This interpretation is counterintuitive in part because of the principle of proximity. We know we're not getting rich in our lifetime if we earn 5 percent annually on our money, so 5 percent looks low. However, we can see just how big it becomes if we let it keep compounding. Note that the overall numbers wouldn't be nearly as dramatic if that rate were cut in half, for example. That explains why even an apparently small percentage contribution of alternative fuels, if enough to suppress the growth rate of oil usage, could extend the life of existing oil supplies by a significant factor.

The issue of population growth involves much smaller percentage figures, but the effect of compounding is no less decisive. To set the stage, note that for the first half a million years of mankind's existence, population growth was near zero. The biggest jump in the growth rate came about during the Industrial Revolution, and that spurt was due to a dramatic lowering of the death rate, not an increase in the birth rate. Growth rates eventually peaked during the 1960s, when the world's population was growing at the rate of 2 percent per year.

By the early nineties that figure had come down a bit, to the 1.7 percent level, and by the turn of the century it had declined to about 1.4 percent. Those declines may not look like much, but growth rates at such a low level are extremely sensitive. In 1993 the population was expected to reach 11 billion people by 2034; with the lower rate, that figure wouldn't be attained until 2043. But the overall growth rates cover up the wide disparity between the developed world and the less-developed world. In the latter, growth rates are still in the torrid zone of 1.9 percent, while population growth within developed countries has all but disappeared.

We recognize that concerns about world population date back to the eighteenth century and Thomas Malthus. The fatal flaw of Malthusian warnings about population growth outstripping resources was that they didn't take productivity improvements into account. While technology is clearly capable of keeping up with population growth for decades or even centuries, it's safe to say that even a 1.4 percent growth rate, sustained over the first two centuries of the new millennium, would pose a serious threat to the quality of life on this planet.

The Rule of 72

When dealing with a growth rate, whether of money or population, the absolute number is often less meaningful than a question we might ask of that number: How long does it take for our money (or our population) to double? The calculation doesn't look friendly at first glance, and indeed to arrive at a precise answer you need to compute logarithms, but there's an easy way out in the form of the so-called Rule of 72. If your money is growing at the rate of 8 percent per year, simply divide 8 into 72 and presto—you will double your money in $^{72}/_8$, or nine years. Logarithms and a few conversions would get you the exact answer of 9 years, 2 days, 8 hours, 39 minutes, and 46 seconds, for those who don't think the easy way out tells you everything you need to know.

An alternative to the Rule of 72 is the Rule of 70, which works exactly the same way. In the fine tradition of taking the easy way out, the decision of whether to use 70 or 72 is usually determined by divisibility. If the growth rate is 6 percent, use the Rule of 72; with 7 percent growth, use the Rule of 70. Conveniently, either 70 or 72 is divisible by every number from 1 to 10. If you're not choosing on the basis of divisibility, we would point out that the Rule of 72 is more suited to annual compounding, whereas the Rule of 70 works nicely for weekly compounding. Also, the Rule of 70 is slightly more accurate for low rates of growth such as population figures, whereas the Rule of 72 becomes more accurate for higher growth rates. If you happen to live in Turkey, where at the time of this writing interest rates were on the order of 34 percent per annum, you might turn to the unheralded Rule of 80 to estimate your doubling time.

Once you've calculated a doubling time, you have to remember that it represents a multiplicative phenomenon, much like the area of a square. Doubling your money over 9 years means quadrupling your money over 18 years, and so on. But the temptation to make things additive (i.e., linear) is always there, especially for intervals shorter than the doubling time. Again, steady 8 percent growth will double your money in 9 years, but your $4\frac{1}{2}$-year return won't be 50 percent. That can't be right because two 50 percent gains would amount to a factor of $1.5 \times 1.5 = 2.25$, not 2. Working backward, we see that the $4\frac{1}{2}$-year return must be worth a factor of the square root of 2, or 1.41: a return of 41 percent. The same reasoning applies to the Richter scale, where we earlier established that an extra point multiplies the strength of the earthquake by 10. But what about an extra half-point? The answer is not half of 10, but the square root of 10, or just over 3. At 7.8 on the Richter scale, the devastating San Francisco earthquake of 1906 released 3.1 times as much energy as the 7.3 magnitude quake that struck Hebgen Lake, Montana, in 1959.[6]

Among other things, the link between growth rates and doubling time allows us to quickly reconcile two frequently cited and apparently conflicting numbers about the risks of tobacco use. Cigarette smoking, it is said, doubles one's chances of dying at any given age. If a 60-year-old nonsmoker faces a 1 percent risk of dying in the following year, then a 60-year-old smoker has a 2 percent mortality risk. Yet it is also widely reported that the life expectancy of a lifelong smoker is around seven years less than for a nonsmoker. That's not even a 10 percent shortening in expected life span. How can that be?

The answer takes us back to 1825, when Englishman Benjamin Gompertz observed that after the age of 20, mortality rates increased exponentially, doubling every seven years. Other things being equal, "Gompertz's Law of Mortality" suggests that a 45 year old is twice as likely to die in the next year as a 38 year old, and a 79 year old faces four times the mortality risk of a 65 year old. Now let's apply Gompertz's Law to smoking. If smoking doubles mortality, then a smoker will face the same mortality risk as a nonsmoker seven years his elder. Hence, smokers will, on average, die about seven years younger than nonsmokers.

Again we have a demonstration of the enormous influence of growth rates over time. Smoking is incredibly unhealthy. To say it once more, smoking *doubles* the chance of death in any year. But because mortality rates increase exponentially with age—by 10 percent per year, according to the Rule of 70—life expectancy is dominated by mortality rates at older ages. Smoking may double one's mortality risk, but being seventy years old instead of thirty-five is 32 times as deadly. Smoking is lethal. Aging is even deadlier.

Exponential Decay

When something increases by a certain percentage in every time period, as when a person's mortality risk rises by 10 percent per year, you have exponential growth. However, quantities can also steadily decline in percentage terms, decreasing, for example, by 10 percent a year, or 50 percent an hour. When that happens, you have exponential decay.

The first thing to notice about exponential decay as against exponential growth is that marginal changes get smaller instead of bigger. A rule of thumb in the used-car world says that cars depreciate by 15 percent per year in personal use and by 20 percent annually in commercial use. Obviously some brands and models hold their value better than others, and depreciation rates will also vary according to conditions of supply and demand in the used-car market. But if you buy a brand-new family car for $20,000, it's reasonable to expect it to lose 15 percent of its value, or $3,000, the first year. The second year the car will again depreciate by 15 percent, but starting from $17,000 the loss is a smaller $2,550. In the third and fourth years of its life, the car will depreciate by $2,168 and then $1,842. When you purchase a used car, not only do you get a lower price, you also miss the steepest part of the depreciation curve.

Recognizing that depreciation often follows the nonlinear pattern of exponential decay, the tax code provides the "declining balance" method of depreciation, which figures depreciation by applying the same depreciation rate each year, just as we did with your hypothetical family car. But there's an interesting wrinkle. Notice that when an amount declines exponentially, it never actually reaches zero. No matter how many times you reduce a quantity by 15 percent, 20 percent, or even 99 percent, there's always something left. So in order to keep the length of depreciation schedules finite and reasonable, the declining balance method switches to "straight line" depreciation—depreciation by a fixed dollar amount each year rather than a fixed percentage—when the straight line method provides an equal or greater deduction. For example, a business depreciating a conference table, which as office furniture is considered seven-year property, might end up using the declining balance method for four years and then the straight line method for three.

We've talked about doubling times with respect to exponential growth, and the corresponding concept in exponential decay is the half-life. You've most likely heard the term in relation to radioactivity, where half-life denotes the time it takes for half of the nuclei in a sample of radioactive material to undergo spontaneous disintegration, thereby emitting radiation. But half-life can be used to describe the rate of any kind of exponential decay. Most drugs, for instance, are eliminated from our bodies at a constant percentage rate per unit of time. The half-life of such a drug is the time required to eliminate half of the amount of the drug present in the body. (Curiously, the exponential breakdown of drugs doesn't apply to alcohol, which is ordinarily metabolized at a linear rate of 10 to 15 ml per hour. In frat-boy math, that's about 1 beer per hour.)

Half-life is often a decisive factor in choosing among alternative drugs. The popular sleep aid Zaleplon (marketed as Sonata) has a half-life of just an hour, which means that in eight hours over 99 percent of it will get eliminated. Because Zaleplon wears off so quickly, it doesn't cause morning drowsiness. The downside of the short half-life is that Zaleplon doesn't increase sleep time. By contrast, Doxylamine (the active ingredient in Unisom) has a half-life of around 10 hours.[7] If you take Doxylamine, it will help you sleep longer, but you may have to get yourself to work in the morning with half of the drug still in your system.

Half-life also helps determine how frequently drugs are taken. Consider the anti-inflammatory drug Aleve, which has a half-life of about 12 hours. That slow rate of metabolism allows for long intervals between dosages. If you take an Aleve pill, half of it will still be in your system 12 hours later. By contrast, Aleve's leading competitor, Advil, typically has a half-life of around 2 hours, so that 12 hours after taking an Advil pill, only 1/64th of it will remain in your body. Therefore, in order to maintain similar pain relief, Advil has to be taken much more frequently than Aleve, apparently enough of an inconvenience that many pain sufferers take Aleve even though it is more than twice as likely as Advil to cause adverse gastrointestinal effects.

In case you were wondering, the Rules of 70 and 72 can be used with half-lives, the idea being that the decay rate times the half-life equals one of those two numbers. But the rules are generally less accurate for half-lives than for doubling time and will always overestimate the half-life or decay rate. With that in mind, you might make a small downward adjustment to whatever answer the rules give you. If a car depreciates by 15 percent a year, the Rule of 70 predicts it will lose half its value in 4 years and 8 months. In fact, the half-life is about 4 years and 3 months.

Curvature in Policy: Progressive Taxation

In the previous chapter we mentioned Mark McGwire's walking away from a $30 million contract, retiring rather than performing below the level he had set for himself. The immediate conclusion was that Mark McGwire has an admirable set of personal standards. The secondary conclusion was that he could afford to be admirable because of the fortune he had already tucked away.

However, there is an underlying point that is far more important than Mark McGwire's character or personal balance sheet. The point is that well-being does not increase *linearly* with wealth, especially at the extremes of the wealth spectrum. This simple observation has important implications in the area of public policy. You begin with the fact that the marginal benefit of an extra $1,000 is much greater for someone earning the minimum wage than it would be to, say, Steve Forbes. What you end up with is progressive taxation.

Progressive taxation stems mainly from the premise that the burden of income taxes should be shared equally among citizens. There's more than one way to interpret the notion of equal burden, but because of the declining marginal value of money, most people would agree that forgoing a third of your income is less of a burden if you earn a million dollars than if you make $10,000. The implication is that a taxpayer with $1 million in income should face a higher average tax rate than a taxpayer earning $10,000. Mathematically, the only way to accomplish such progressive taxation is to have more than one marginal tax rate.

There's a lot of confusion about whether or not a "flat tax" would be progressive. "Flat" doesn't sound progressive, and flat-tax advocates emphasize the simplicity and apparent fairness of a single, uniform marginal tax rate. In fact, flat-tax schemes, because they include a sizable exemption, actually have two marginal rates. Under the most prominent proposal, advanced by Stanford professors Robert Hall and Alvin Rabushka in their book *The Flat Tax*, a family of four would face a marginal tax rate of 19 percent on taxable income above an exemption of $25,500 (indexed to inflation).[8] In other words, there are two marginal rates: 0 percent on the first $25,500 in income, and 19 percent on additional earnings.

On that score, the flat tax is progressive, especially at lower levels of income. A family earning $30,000 would pay 2.9 percent of its income in taxes, while a family earning $60,000 would pay an average rate of 10.9 percent. But there's a big, and often overlooked, catch. Back in the third chapter we emphasized that you should always ask how things are defined and measured, and in this case you need to ask those questions about taxable income. What you'll discover is that interest, dividends, and capital gains are not taxed under the Hall/Rabushka plan. Since such investment income accounts for an especially large share of the income (as traditionally defined) of many wealthy individuals, flat-tax schemes that exempt investment income are probably regressive at higher income levels.

The Curvature of the Flat Tax

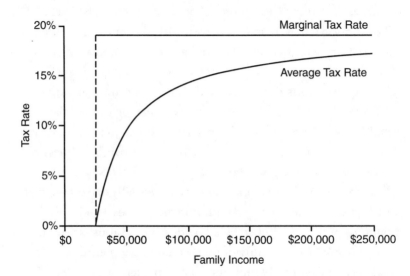

In fact, when you ask questions of the Hall/Rabushka flat tax, you'll come to the surprising realization that it's not even an income tax at all. All business investment is treated as an immediate expense (there is no depreciation), thereby removing investment from the tax base. Because income equals consumption plus investment (the income you don't spend gets saved and invested), the Hall/Rabushka flat tax is actually a consumption tax that on its surface looks like an income tax. Granted, there are strong arguments for taxing consumption instead of income, and it's worth emphasizing that progressivity is only one of many considerations that should influence the design of a tax system. But the fact is that consumption taxes tend to be regressive because saving rates usually rise with income levels.

Ponzi Schemes, Savings and Loans, and Social Security

The name "Ponzi scheme" has a colorful history. The initial such scheme, concocted by Boston entrepreneur Charles Ponzi, actually began as a business with a thread of legitimacy. The business revolved around International Postal Reply Coupons, which were introduced in 1907 following a wave of immigration into the United States. By inserting a coupon into a letter, a sender of overseas mail could provide for return postage; the recipient could redeem the coupon for stamps denominated in the recipient's home currency. Following World War I, many European currencies weakened, creating some wild imbalances. You might, for example, purchase a coupon in Spain for the equivalent of one American cent, only to find that you could redeem that coupon for five or six cents when the letter reached the United States. Ponzi sought to exploit these price differences in a formal way, and he established a company to do just that. (Ironically, he dubbed his venture the Securities Exchange Company, similar to our current governing body, the SEC.) Ponzi promised extraordinary payouts to his investors, as in doubling their money in 45 days.

The good news was that money flooded in from the investing public. The bad news was that even if Ponzi had wanted to conduct a legitimate enterprise, there simply weren't enough International Postal Reply Coupons outstanding to keep the business going, certainly not enough to match the money his organization was taking in. So when the 45-day period for one batch of investors came up, the only way for Ponzi to keep his promise was to pay them off with the proceeds from subsequent investors.

Ponzi schemes, in case you didn't notice already, involve exponential growth. The reason we wanted to talk about Ponzi schemes in this chapter is that they are highly nonlinear. Specifically, Ponzi schemes involve the kind of exponential growth we just discussed. To double your money is one thing. To do so on the basis of contributions from the next generation of investors is equivalent to saying that each subsequent generation must be twice as numerous as its predecessor. If you started with 100 investors, in ten generations you'd need to hook 102,400 patsies into your scheme, all without gaining the attention of the authorities. Good luck.

What makes the original Ponzi story so remarkable is that he gained support long after the quantitative lunacy of his scheme had been exposed. Such was the power of word of mouth from that happy first generation of investors. Somehow no one thought to check on just how many postal coupons were actually redeemed. When that task was eventually completed, the disparity wasn't hard to see. Whereas Ponzi would have had to redeem at least 180 million coupons to come anywhere close to the requirements of his alleged business, the actual number of redeemed coupons was . . . two.

It is illegal to run Ponzi schemes, but it isn't illegal to run unbalanced businesses. Today, when we think of the savings and loan crisis, we think of Charles Keating and other dubious characters. But the *first* S&L crisis arose from a failure to plan for a change in curvature. The fundamental business of a savings and loan institution is to lend out money in long-term contracts, as in 30-year home mortgages. The money, historically, was obtained by borrowing at short-term rates. You make money as long as long-term rates exceed short-term rates, which people figured would always be the case. Beginning in late 1978, however, the nation experienced a so-called "inverted yield curve." With long-term rates lower than short-term rates, a typical S&L was now losing money every time it made a long-term loan. A long list of bankruptcies ensued in the early 1980s. Those institutions that did make it through—notably federal agencies Freddie Mac and Fannie Mae—did so by "balancing" their portfolios through a process called duration matching. The particulars aren't important (well, they are, but they're complicated and beyond our current scope), but the important result of this rebalancing is that the agencies were prepared for yield-curve fluctuations.

These days, you're most likely to hear the term "Ponzi scheme" if you listen to certain critics of Social Security. That's because each generation of Social Security participants gets paid off with the contributions of the next generation. But is this really a Ponzi scheme? At first glance the system looks unsustainable. The ratio of workers (who contribute to Social Security) to retirees (who receive benefits) has steadily declined, forcing substantial increases in payroll tax rates in order to pay promised benefits. Whereas the original contributions to Social Security took 2 percent of a typical worker's pay, today the payroll tax rate stands at 12.4 percent. And the future looks set up for more of the same: on intermediate projections, the worker-to-retiree ratio is expected to fall from 3.4 to 2.0 over the first half of the twenty-first century.[9]

Yet unlike a classic Ponzi scheme, Social Security is not destined to collapse. Remember one of the lessons from our discussion a few pages back: Rates of return are inherently sensitive numbers. When workers pay Social Security taxes and later collect benefits, they have effectively earned some rate of return on their contributions. The greater the benefits are relative to past contributions, the higher this implicit rate of return. The sustainability of a pay-as-you-go pension arrangement such as Social

Security depends on this rate. If set unrealistically high, the transfers from workers to retirees will require ever-increasing tax rates and the Ponzi charge has some merit. But there's always some rate of return that is economically sustainable, although it may require substantial cutbacks in future benefits. If Ponzi had offered his investors 1 percent a year, his program could have been fiscally sound, although it wouldn't have attracted much interest. A central political challenge for Social Security is whether its popularity will wane if and when benefits are reduced.

When you look at the numbers, it appears inevitable that the implicit rate of return on Social Security contributions has to fall, presumably through some combination of higher payroll taxes, lower benefits, and an older retirement age. Despite the rhetoric of some politicians, there is no painless solution to the generational challenges the system now poses. Good quantitative thinkers understand that where curvature is concerned, there's not always an easy way out.

Taking Chances

In the previous six chapters, we explained how grammar school and high school math classes fail to teach many critical quantitative skills, and we began filling in that gaping educational hole by suggesting a host of new ways to look at, think about, and work with numbers. As if that undertaking weren't ambitious enough, in this chapter we are going to tell you what life is all about, and then we'll help you cope with that disclosure.

So what is life all about? Recognizing that this question has already tortured souls for millennia, we won't delay the answer with pages of philosophical fluff. *Life is about making decisions under uncertainty.* Like: When should I leave for the office in the A.M.? Should my company increase its advertising budget? Should I order the veal marsala or the chicken cacciatore? Should I have my girlfriend's name tattooed on my bicep?

Life is characterized by constant decision-making, and in every case uncertainty dominates the considerations. How bad will the traffic be? How much revenue will each additional dollar of advertising generate? Will the chicken be overcooked? Will my girlfriend see the tattoo as a sign of my commitment, or as conclusive evidence that I haven't a shred of common sense?

There is more than one possible answer to all such questions, and the correct answers can never be forecast with certainty. Thus, to quote former treasury secretary Robert Rubin again, "All decisions become matters of judging the probability of different outcomes." If I leave the house at 8 o'clock in the morning, what's the chance it will take 20–30 minutes to drive to the office? 30–40 minutes? An hour? What if I leave at 7:30? Only when these probabilistic questions have been answered satisfactorily can one make an informed decision about when to depart.

To be sure, decisions are also based on values. If someone doesn't believe in punctuality, then even clairvoyant traffic reports wouldn't get him to work on time. But the relevance of values doesn't change the fact that even the weightiest of decisions call for forecasts of the likelihood of different outcomes. During the 2000 presidential campaign, Al Gore said that whenever faced with a difficult decision, he asks himself, "What would Jesus do?" But when, in June 1945, President Truman was considering how to bring the war with Japan to a close, a deliberative process that led to his decision to use atomic weapons, he didn't start by consulting a moral philosopher or religious prophet. Instead, he requested numerical forecasts that would help clarify the uncertainty of the situation. At Truman's behest, Adm. William Leahy sent an urgent memorandum to the other members of the Joint Chiefs of Staff, stating that the president wanted "an estimate of the time required and an estimate of the losses in killed and wounded that will result from an invasion of Japan proper. He wants an estimate of the time and the losses that will result from an effort to defeat Japan by isolation, blockade, and bombardment by sea and air forces."[1] Truman didn't ask what Jesus would do. Truman's real question was, "What are the odds?"

What Is Probability?

When you make decisions under uncertainty, you're making bets. And if you're lousy at estimating the relevant probabilities, you're destined to make a lot of bad bets. Indeed, your best opportunity for improving your decision-making may be to improve your skill at sizing up probabilities.

The first step to doing this is to understand what probability is. You may think you already do—isn't probability simply the likelihood that an event will happen? Well, yes, but that begs the question of what "likelihood" means. Probability is an elusive concept. In the pages that follow we will introduce you to the three leading interpretations of probability—classical, frequentist, and subjectivist—which mathematicians argue for and against with religious fervor. God help the "subjectivist" assistant professor trying to get tenure in a statistics department ruled by dogmatic "frequentists." Happily, we don't have to take sides in this never-ending debate. To be a skilled quantitative thinker, you need to be aware of all three approaches, appreciating their respective virtues and limitations.

Continuing with the theme of decision-making as betting, consider roulette, the oldest casino game still in operation. In American-style roulette the wheel has 38 slots, numbered 1 to 36, plus 0 and 00. (European wheels normally lack the 00 and thus have 37 slots.) Of the slots numbered 1 to 36, half are black and half are red. The 0 and 00 slots are green. Now suppose you wager that the ball will come to rest in a red slot. What is your probability of winning?

Obviously the house must have an advantage, so your chance of winning has to be less than 50 percent. (Blaise Pascal, the seventeenth-century French philosopher and mathematician credited with inventing roulette, was also the father of modern probability theory, so he wasn't about to miss that nuance.) To get the exact figure, note that there are 38 possible outcomes—what mathematicians call the sample space—18 of which are favorable. The probability of winning is simply $^{18}/_{38}$, or 0.474. This ratio is an illustration of what is termed the "classical" interpretation of

probability: The probability of an event equals the number of favorable outcomes divided by the number of possible outcomes, assuming they are all equally likely.

Notice that we calculated a probability of 0.474 without
reference to any data on actual spins of a roulette wheel. The
classical definition of probability is a theoretical one. The advantage of this approach is that classical probability asks you to
set aside data and think hard about the nature of an event. *In
theory*, what is the probability the ball will stop on a red slot?
In theory, what is the chance that stocks will outperform bonds
over the next decade?

Another benefit of the classical view of probability is that it
requires you to figure out all of the ways in which the event in
question can occur, or, put another way, what the sample space
looks like. The sample space in roulette is staring us in the face,
but it's not always that simple. Many sample spaces are quite a
bit less concrete and turn out to be much bigger than we'd ever
imagine.

The attention accorded so-called Bible codes in the late 1990s
was the result of a monumental misunderstanding about sample
space sizes. According to believers, the Torah—the first five books
of the Hebrew Scriptures—contains hidden messages about the
future. The prevailing method for uncovering these alleged
prophecies is to search the text by examining equidistant letter
sequences (ELSs). ELSs are created by starting with a given letter
in a text and then repeatedly skipping (forward or backward) a
certain number of letters, ignoring spaces and punctuation. Findings are often displayed by arraying the letters of the text in a grid
where the row length equals the skip code. ELSs then show up
vertically in the grid. As an illustration, consider the sentence,
"There are messages about Bible codes that are hidden in this
text," arrayed with a skip code of 14.

```
T  H  E  R  E  A  R  E  M  E  S  S  A  G
E  S  A  B  O  U  T  B  I  B  L  E  C  O
D  E  S  T  H  A  T  A  R  E  H  I  D  D
E  N  I  N  T  H  I  S  T  E  X  T
```

Then again, those who doubt divine intervention might prefer a skip code of 17.

```
T  H  E  R  E  A  R  E  M  E  S  S  A  G  E  S  A
B  O  U  T  B  I  B  L  E  C  O  D  E  S  T  H  A
T  A  R  E  H  I  D  D  E  N  I  N  T  H  I  S  T
E  X  T
```

Bible codes have had adherents since the 1940s, but it was Michael Drosnin's 1997 book, *The Bible Code*, that brought the subject to a popular audience. Drosnin reported countless seemingly portentous ELSs in the Hebrew Bible, the most notable of which was an ostensible foretelling of the assassination of Israeli prime minister Yitzhak Rabin. The uncovered phrases are certainly eye-catching, and more than a few people, including some we know, have accepted Bible codes as conclusive proof, not only of the divine authorship of the Bible but also of the view that the Bible can be used to predict the future.

But before you conclude that Bible code messages should be placed in fortune cookies, it's worth looking at ELSs from the perspective of classical probability. What we discover is that as the size of a document increases, the total number of ELSs blows up to unfathomable proportions. That's because each skip code (except for the really big ones) can be applied from tens or hundreds of thousands of possible starting points (the Torah has about 300,000 letters). And the skip codes themselves get ridiculously big. Drosnin, for example, skipped 4,771 letters at a time to find the Hebrew equivalent of "Yitzhak Rabin," which is another way of saying that the first 4,770 tries didn't work.

Take a 250-page book with 500,000 letters. The book will contain over half a trillion ELSs of between three and twelve letters in length—about 125 billion three-letter ELSs, 83 billion four-letter sequences, 62 billion five-letter strings, and so on. Once we remind ourselves just how big half a trillion really is, it's obvious that such a humongous sample will include millions of words, not by virtue of the Almighty, but by pure chance. Assume, for the sake of simplicity, that all letters are equally likely to show up in this hypothetical book. We would then expect the word "kill" to occur in one out of every 456,976 four-letter sequences,* or over 180,000 times altogether, often in close proximity to names like "Bush" and "Gore," which would both show up with similar frequency. "God" would likely appear over 7 million times. And note that Drosnin didn't conduct a search for a few specific words or names selected in advance. He cast a much wider net by choosing words and names only after they happened to turn up, presumably ignoring vocabulary that didn't suit his purposes. That's like shooting fish in a barrel.

This last point—that specific matches are considerably less likely than general matches—is known to anyone who has ever conducted a web search. We did a Google search for "Joel Best" (the author of the book *Damned Lies and Statistics*) and got 3,100

---

* With 26 letters in the alphabet, there are $26 \times 26 \times 26 \times 26$ possible four-letter sequences.

matches, having used quotation marks to guarantee that the referenced web sites contained his full name (or anyone else by that name, but we couldn't stop that). Our next search omitted the quotation marks, so we got a listing of web pages that contained a "Joel" and a "Best" anywhere at all. By the time you add up the sites that mention Joel Grey's Oscar for Best Supporting Actor, Haley Joel Osment's nomination within that same category, the CD *The Best of Billy Joel*, and everything else, you're talking about 1,030,000 web pages. The point is that the "messages" found in *The Bible Code* were humdrum general matches masquerading as rare specific ones. Who's to say that Drosnin's technique didn't uncover the phrase "Rabin lives"?

In response to critics who claimed his discoveries were the unremarkable product of chance, Drosnin responded, "When my critics find a message about the assassination of a prime minister encrypted in *Moby Dick*, I'll believe them." Famous last words. Australian mathematics professor Brendan McKay promptly took up the challenge and found in Melville's book apparent mentions of the murders of Rabin, Indira Gandhi, Abraham Lincoln, and John Kennedy, among others, including, in a delightful coup de grâce, Drosnin himself.

Of course, some letters occur more frequently than others. *E*'s, for example, are more common than *u*'s. And so a search of eight-letter ELSs generated from an English-language text, looking for the names of great track-and-field athletes, would be far more likely to encounter discus legend "Al Oerter" than middle-distance star "Qu Yunxia." And therein lies a major limitation of classical probability. The classical interpretation of probability assumes that all outcomes are equally likely, which is often not the case. In fact, probabilities are unequal more often than you may realize.

The Frequentist School

In 1662, London haberdasher John Graunt published a short book with a long title. Graunt is widely credited as history's first epidemiologist and demographer, and he summarized his pioneering research in *Natural and Political Observations Mentioned in a following Index, and made upon the Bills of Mortality.* Included in this book was the curious observation that between 1629 and 1661, 139,782 males and 130,866 females were christened in London parishes of the Anglican Church.[2] Put in contemporary terms, Graunt determined that the "sex ratio" at christening was 1.068. That is, there were 1,068 males for every 1,000 females.

Crossing the Channel and fast-forwarding a century, we find French mathematician Pierre Simon Laplace tallying hundreds of thousands of births throughout France as well as in London, St. Petersburg, and Berlin. Laplace concluded that the ratio of male to female births was approximately 22 to 21, a sex ratio of 1.047.[3] Today it is assumed that the biologically normal sex ratio at birth is 1.06, not the straight 50–50 shot that we might have expected.

Collecting data on boys and girls is an illustration of the "frequentist" approach to probability. Here the probability of an event is defined as its relative frequency in a large number of similar trials. What then is the probability that a newborn baby is a boy? The long-run ratio of male births to female births is 1.06, which means there are 106 boys for every 100 girls, or 106 boys for every 206 births. Thus the probability of a boy is $^{106}/_{206}$, or 0.515.

This calculation seems straightforward. Yet many find the frequentist conception of probability unsatisfactory. A major problem is interpreting what a long-run average signifies in the case of a single event, as when a weatherman says there's a 10 percent chance of precipitation tomorrow. ("Precipitation," to a meteorologist, means at least 0.01" of liquid precipitation.) It's either going to rain tomorrow or it isn't. As with pregnancy, there's no middle ground.

The frequentist reply is awkward. What the weatherman is really saying is: "In the long run, I expect it will rain on about 10 percent of days following days where those meteorological conditions that I can observe appear similar to what I'm seeing now." Some find this mouthful acceptable; others think it merely begs the question of what it means to say that there's a 10 percent chance of precipitation within a particular area and time frame. No surprise that it can be hard for people who have different perspectives on probability to have a useful conversation, as one of us was reminded when his wife became pregnant during the writing of this book.

(The curtain rises on the dining area of a loft-style apartment in a converted warehouse. Odds-playing, frequentist-minded HUSBAND sits at the head of a narrow, weathered, olive wood dining table with expecting WIFE seated to his right.)

WIFE: What do you think it is? A boy or a girl?

HUSBAND: I think there's a fifty one and a half percent chance it's a boy.

WIFE (sighing and rolling her eyes): I know the statistics—you've told me them before. But the baby's sex is already determined and I'm asking what you think it is.

HUSBAND: I think it's either a boy or a girl, and that it's more likely it's a boy. If you want me to guess, obviously I'm going to guess it's a boy.

WIFE (heading for the kitchen): You're impossible.

In the event, the baby was a girl, which proves . . . absolutely nothing.

Frequentist reasoning can serve as a real-world safeguard for some of the traps of classical probability. Consider the diagram below, which depicts the playing surface of an old amusement park game. To play the game you throw nickels (yes, your nickels) onto a table patterned with colored-in circles and lipped so that all tosses stay within the playing surface. If the nickel comes to rest *entirely* within one of the circles, you win your choice of Kewpie dolls. If not, all you get is the experience of literally throwing your money away.

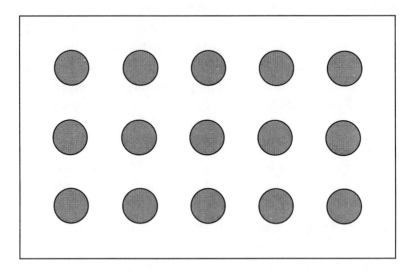

The probability of winning the game appears to be the ratio of the total area of the shaded circles (favorable outcomes) divided by the area of the rectangle (total sample space). To the naked eye, those odds don't look all that bad, and the preliminary arithmetic confirms that view: To make the ratio as simple as possible to calculate, we'll inflate our currency just a tad and suppose that a nickel is an inch in diameter; in addition, we suppose that all other relevant lengths (the diameter of each circle, the distance between adjacent circles, and the distance between the outer edge of any perimeter circle and the edge of the table itself) are precisely two inches. The total shaded area is therefore 15π (each circle has radius 1 and area π square inches), and the area of the rectangle is $14 \times 22 = 308$ square inches. The apparent "winning probability" is thus equal to $15\pi/308$. That's 15.3 percent, or slightly better than a one-in-seven shot.

Buoyed by the reasonable-looking odds, the typical sucker

starts pitching nickels onto the table, only to find out that the game is much tougher than it looks. This frequentist approach soon suggests that the actual probability of winning is substantially lower than one in seven.

Did classical probability break down? No, but our arithmetic sure did. The problem is that the game is an optical illusion, because the shaded areas do not in fact represent "winning" regions. To see why, you have to realize that in order to win the game, the center of the nickel must come to rest within an imaginary smaller circle nested inside one of the larger ones. The radius of one of those smaller circles is half an inch, so its area is $\pi \times (\frac{1}{2})$ squared = $\frac{\pi}{4}$ square inches. There are 15 circles altogether, so the total "winning area" for the center of the nickel is $\frac{15\pi}{4}$ square inches. On the other hand, the total area of the rectangle in which the center of the nickel might fall is 21×13, or 273 square inches. (The entire board is 22×14, but the center cannot possibly land within $\frac{1}{2}''$ of the border, as indicated by the dotted line.) The actual probability of success is therefore 15π divided by 4×273, only 4.3 percent!

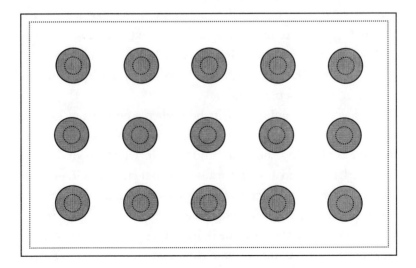

This illusion is quite tricky, and similar ones have ensnared even those who truly live by their wits. At the beginning of the 2001 baseball season, full-time baseball fan and part-time political observer George Will commented in the *Washington Post* on the lamentable irregularities in the major league strike zone. A reasonable complaint, but here's the way he started:

> Through most of the 20th century the strike zone was defined as a rectangle extending over the 17-inch-wide plate, from the batter's shoulders to his knees. Actually, the rectangle is about 22 inches wide, because a pitch is a strike if any part of the ball crosses the plate, the width of which thus is, effectively, 17 inches plus the width of two baseballs.[4]

Close, but not quite. Unless you want to relive the amusement park deception, you must define the strike zone in terms of a single point—the center of the baseball. The width of the effective strike zone for that single point must be 17 inches plus the width of *one* baseball (one half on either side), because that's the width of the space into which a pitcher can throw the center of the baseball and still get a strike. Giving a pitcher an extra two-and-a-half inches isn't such a small thing, as Will himself would acknowledge. In his defense, though, we have yet to see him pitching nickels at an amusement park.

Frequentist reasoning is at its best when a prospective event is analogous to past events for which there are ample data. We know that the probability the next child born in Dubuque, Iowa, will be a girl is around 0.485 because researchers like Graunt and Laplace have literally counted millions of births. In turn, we know something is wrong when the corresponding figure from a Chinese census turns out to be 0.461.[5] It is assumed that the discrepancy is attributable to some combination of the underreporting of female births, excess female infant mortality (at least partly due to infanticide), and sex-selective abortion of female fetuses.

On a more optimistic note, frequentist thinking reminds us that the stock market offers investors a built-in advantage. Many investors wrongly dismiss the stock market as a coin toss, in which the likelihood of the market going up on any given day is 50 percent. But the efforts of researchers such as Salomon Smith Barney's Louise Yamada have shown that the frequency of up days has been 52.6 percent over a century's worth of data. For once the little guy owns the house advantage, a boon to small investors everywhere.

However, limits on frequentist thinking arise because most events are, if not unique, sufficiently rare or idiosyncratic that any seemingly relevant historical data must be interpreted with care. Put yourself in the shoes of an actuary asked, on August 12, 2001, to estimate the likelihood that Chicago would in the next year suffer a fire on the scale of the Great Fire of 1871, which decimated some 18,000 buildings over a 2,000-acre area. That such a conflagration had occurred once in the 168 years since the Town of Chicago was incorporated (it became a city four years later) would not make $\frac{1}{168}$ a sensible estimate. Cows may be as careless as ever, but like the women in Virginia Slims ads, building construction and firefighting have come a long way.

And so, for the most part, have insurance companies, although prior to September 11, 2001, they engaged in shortsighted frequentist thinking about the risk of significant terrorist-related losses. In a letter to his Berkshire Hathaway shareholders dated November 9, 2001, Warren Buffett admitted that the insurance industry had fallen for the trap of thinking that because something had never happened, it never would:

> A mega-catastrophe is no surprise: One will occur from
> time to time, and this will not be our last. We did not,

however, price for *man-made* mega-cats, and we were foolish in not doing so. In effect, we, and the rest of the industry, included coverage for terrorist acts in policies covering other risks—and received no additional premium for doing so. That was a huge mistake and one that I myself allowed.[6]

The flip side of this mistake is that once a major terrorist act did occur, a frequentist perspective tended to overstate the likelihood of another such man-made disaster. In the months following the World Trade Center tragedy, insurers found themselves in the peculiar position of raising rates and experiencing higher demand in the process. Amid this anomaly, Lloyd's of London was admonished for its insensitivity in dubbing the period "a historic opportunity," but they had a point: In many cases insurers could limit or even leave out their liability for terrorist actions. But the bottom line is that frequentist thinking has real trouble coping with events of extremely low likelihood, which leads us to our third and final approach to probability.

Subjective Probability

Now suppose the actuary assessing the one-year Chicago Great Fire risk had set the odds at 1 in 200. Was he wrong? The actuary might respond, "No, I was right. I said there was a 99.5 percent chance that Chicago wouldn't go down in flames, and it didn't." Now imagine that another actuary had guessed a Great Fire risk of 1 in 100,000. Which estimate was better? On what basis could we say? And if probability forecasts can't be judged right or wrong, or at least better or worse, what use are they?

Such difficulties have led many to the "subjectivist" conception of probability. On this view, probability estimates represent personal beliefs. When an economist says there's a 30 percent chance of a recession next year, he is expressing a level of confidence in his judgment that the economy won't contract for two consecutive quarters (the customary definition of a recession). He's 70 percent sure there won't be a recession.

Subjectivists make an important point when they stress that most probability estimates involve personal judgment. But just because reasonable people may disagree about the likelihood of an event does not imply, as some assert, that all probability estimates are equally reasonable. Arguing that a climatologist has overestimated the likelihood that the average surface temperature of the earth will rise by more than 2°C over the next century is not akin to saying that Coke tastes better than Pepsi. Oddsmaking is not a matter of taste; probability estimates are a matter of reasoning and calculation, which can be done well or poorly.

However, because of the influence of chance, probability estimates cannot be judged solely on the basis of outcomes. When a patient survives cancer after his oncologist estimates only a 10 percent chance of recovery, the oncologist appears to have been mistaken. But remember, if the oncologist's oddsmaking is on target, he will be "wrong" with 10 percent of such patients.

For the most part, probability estimates need to be judged on the quality of the reasoning that generated them. Ouija boards, crystal balls, and tarot cards do not qualify as reasoning. Whether hunches should be considered reasoning is debatable but beside the point. Gut feelings are undependable, and reliance on them tends to be an excuse for avoiding the effort of diligent research and hard thought. In our experience the best oddsmakers try to ignore—if not repress—their intuition.

What skilled quantitative thinkers do in lieu of guessing is to combine classical, frequentist, and subjectivist perspectives. They think carefully about the various ways in which events can play out, they look for analogous historical data, and they ask themselves why that historical data might or might not be relevant. That last step is especially important. It is critical to figure out which data are most germane, but even then you need to appreciate the limitations of those data.

Take an investor trying to forecast the probability of various stock market returns over the next five years. The investor might observe that when the price-to-earnings (P/E) ratio of the stock market is high in historical terms, subsequent returns have tended to be mediocre. However, finance professor Jeremy Siegel, author of the best-selling *Stocks for the Long Run*, counsels that this use of historical data is too general. According to Siegel's research, you have to know *why* the P/E ratio is high: Is it because stock prices are rising or because earnings are falling? Siegel nicely sums up the difference between the two cases: "If the peaks in the ratio were caused by surging prices, that did portend bad stock returns in the future. Five years later, the after-inflation returns were a measly 1.1% per year. But when collapsing earnings caused the peaks of the P/E ratio, the opposite occurred: Real returns averaged nearly 10% per year over the next five years."[7]

Yet even when data are made more specific, rarely will they perfectly fit the circumstances you are trying to evaluate. The best data cannot replace personal judgment, which is why subjectivist reasoning is a needed companion to frequentist analysis.

In 2001 the National Heart, Lung, and Blood Institute issued new guidelines for the treatment of high blood cholesterol in adults. Therapeutic recommendations, such as the use of cholesterol-lowering drugs, are based heavily on a person's 10-year risk of coronary heart disease (CHD). For example, according to the guidelines, drug therapy should be considered for those whose 10-year CHD risk is 10–20 percent and whose LDL cholesterol level is greater than or equal to 130. Ten-year CHD risk is calculated using something called Framingham risk scoring, which assigns people a numerical risk level based on the following factors: age, total cholesterol, HDL cholesterol, systolic blood pressure, treatment for hypertension, and cigarette smoking. A nonsmoking man, age 56, with total cholesterol of 223, HDL of 44, and untreated systolic BP of 125 would have a Framingham point score of 12, which translates into a ten-year risk of 10 percent, right at an important cutoff for considering drug therapy under the guidelines.

But remember one of our habits for dealing with uncertainty: Know what you know and don't know. A lot of information goes into Framingham point scoring, but as a background check it is far from complete. What if a patient has no family history of premature CHD? Wouldn't he have a lower CHD risk than another man with the same Framingham score whose parents both had heart disease? And wouldn't a marathon-running patient have a lower CHD risk than a couch potato with the same Framingham score? A good physician-oddsmaker would adjust for these and other factors not weighed by the Framingham formula. Where probabilities are concerned, subjectivism is just a fancy name for common sense.

Expressing Probabilities

Before we say more about estimating probabilities, let's take a brief detour to discuss how probabilities are expressed. In the course of your life, you will have to communicate probabilities to others and you will have to make sense of other people's probability estimates. There are different ways to express probabilities, and the result is a surprising number of ambiguities and misunderstandings.

Probabilities range from 0 to 1. Zero means no chance. The probability that Robert Banks Davidson will be elected to the

Baseball Hall of Fame is 0. (Don't know who Bob Davidson is? Perhaps that's because his entire major league career consisted of pitching one inning for the New York Yankees against the Kansas City Royals on July 15, 1989.[8] He gave up two earned runs, so his career ERA was 18.00.) A probability of one means an event will occur with certainty. The probability that Bob Davidson will not make it to Cooperstown is 1.

Probabilities are often expressed as percentages, which creates a range of 0% to 100%. Other numerical scales are also used. On his political talk show *The McLaughlin Group*, host John McLaughlin multiplies the traditional 0 to 1 range by ten. For instance, in the first week of George W. Bush's presidency, McLaughlin asked: "On a probability scale of zero to 10, zero meaning zero probability, 10 meaning metaphysical certitude, what's the probability of Bush's getting vouchers through the Congress?"[9]

We shouldn't have to say this, but different scales don't mix. Sometimes scale mixing is obvious, as when Air Force football coach Fisher DeBerry said, "Ninety-nine times out of 10 you're not going to win like that."[10] But when a number is not obviously out of bounds, scale mixing is hard to spot. For example, shortly after John F. Kennedy Jr. died while piloting his single-engine Piper Saratoga, the director of the Aircraft Owners and Pilots Association wrote to *Slate*, defending the safety of general (noncommercial) aviation. The letter noted: "In 1998 there were 361 fatal accidents out of 39 million flights. The 'risk' of a fatal accident is only 0.000009 percent."[11] Oops. When we divide 361 by 39 million, we get a probability of 0.000009, but that's equal to 0.0009 percent, with only three zeros after the decimal point instead of five. Small numbers tend to obscure such arithmetic errors, and unless you're an aviation safety expert, you'd be hard pressed to notice that the figure in the letter understated the risk of general aviation by a factor of 100.

Non-numerical probability scales are common. We've all seen

surveys where the possible answers—such as how likely you are to buy a particular product again—are organized like this:

> Definitely No
> Very Unlikely
> Somewhat Unlikely
> Somewhat Likely
> Very Likely
> Definitely Yes

Whether such verbal definitions yield better responses than numerical estimates is arguable. On the one hand, respondents may have different views about what such terms represent. One person might think of a 20 percent chance as "very unlikely," while another would define it as "somewhat unlikely." On the other hand, many people have a hard time quantifying probabilities and find that verbal descriptions better fit their thinking.

Good quantitative thinkers prefer numbers to phrases, and not just because numbers are descriptively more precise. Numerical estimates also promote sharper thinking than do verbal estimates. If we heard that the chance of something was 60 percent, we couldn't help but wonder if and why 0.60 was a better estimate than 0.55 or 0.65, whereas the word "likely" would prompt no such reflection. When good quantitative thinkers do use verbal descriptions of probabilities, they opt for phrases whose meaning is clear. It's better to say you regard a potential outcome as "very likely" than to say you'd bet "dollars to doughnuts" on its occurrence. The meaning of the often-used latter expression would seem to change with the prices at Krispy Kreme and made a lot more sense when doughnuts cost a few pennies.

Probabilities are often expressed as odds. Odds look harmless enough—just a pair of numbers separated by a dash or colon—but the confusion and obfuscation that go with them are remarkable. If Congress ever decided to act in the public interest, it could do worse than to pass a law banning the use of odds as a method for stating probabilities.

First of all, odds present a kind of optical illusion, causing probabilities shown as odds to be incorrectly perceived as fractions. Odds of 5–2 or 2–5 look like a probability of $\frac{2}{5}$, and not like probabilities of $\frac{5}{7}$ and $\frac{2}{7}$, which is what the odds actually represent. Two-fifths is wrong because odds do not directly show the probability of an event. As one of our dictionaries explains, odds express "the ratio of the probability of an event's occurring to the probability of its not occurring."[12] In other words, odds compare the frequency with which an event occurs to the frequency with which it doesn't occur. That makes odds useful for expressing *relative* probabilities, such as how much more likely a smoker is to die from coronary heart disease than a nonsmoker, but confusing when used to express *absolute* probabilities, such as the chance that a smoker will die of lung cancer.

If the odds on an event are 5–2, then for every five times the event happens, we would expect it not to happen twice. Five out of seven makes a probability of $\frac{5}{7}$. Right? Sometimes. Despite what dictionaries indicate, odds often state the odds *against* something. If the odds on your favorite football team winning its next game are 2–1, that ordinarily implies that there's a two out of three chance your team will *lose*.

The ambiguity about what odds represent leads to even further doubt when people describe odds as changing. When the odds on something go up, does that make it more likely or less likely? It sounds like it's more likely, but when odds show the odds *against* an event, bigger numbers mean a lower probability. In common practice, 4–1 is less probable than 3–1. If you're confused, don't worry, for even if you understand how odds work, you can never be sure if the person you're talking to does. When someone tells you that Tiger Woods had a subpar round, is that good or bad? If someone tells you that the Amazon ranking for *What the Numbers Say* has gone up, does that mean the ranking has gotten better, or does it mean that the number has gone up, in which case the ranking has gotten worse? When someone tells you that the odds on something have risen or fallen, shortened or lengthened, don't assume you know what they're talking about unless it's entirely clear from the context.

It's hardly surprising that odds are so widely used in gambling. By disguising the underlying probabilities, odds also camouflage the spreads through which bookies make their money. Shortly before the 2001–2 NFL season, Ladbrokes, the famous London bookmaker, offered the following odds on the eventual champion. Note that we've displayed the odds with their equivalent probabilities, rounded to four digits.

PRESEASON ODDS ON THE 2002 SUPER BOWL CHAMPION

Team	Odds	Probability
St. Louis Rams	9–2	0.1818
Baltimore Ravens	7–1	0.1250
Tennessee Titans	8–1	0.1111
Tampa Bay Buccaneers	9–1	0.1000
Denver Broncos	10–1	0.0909
Indianapolis Colts	10–1	0.0909
Oakland Raiders	10–1	0.0909
Minnesota Vikings	14–1	0.0667
New Orleans Saints	16–1	0.0588
New York Giants	16–1	0.0588
Green Bay Packers	20–1	0.0476
Jacksonville Jaguars	20–1	0.0476
Washington Redskins	20–1	0.0476
Buffalo Bills	25–1	0.0385

Miami Dolphins	25–1	0.0385
Philadelphia Eagles	25–1	0.0385
Detroit Lions	33–1	0.0294
New York Jets	33–1	0.0294
Kansas City Chiefs	40–1	0.0244
Pittsburgh Steelers	40–1	0.0244
San Francisco 49ers	40–1	0.0244
Seattle Seahawks	40–1	0.0244
Carolina Panthers	66–1	0.0149
Chicago Bears	66–1	0.0149
Dallas Cowboys	66–1	0.0149
New England Patriots	66–1	0.0149
Atlanta Falcons	80–1	0.0123
San Diego Chargers	80–1	0.0123
Arizona Cardinals	100–1	0.0099
Cincinnati Bengals	150–1	0.0066
Cleveland Browns	200–1	0.0050

In gambling, odds of X–Y mean that a successful bet of the amount Y creates a net profit of the amount X, so a £1 bet on the Patriots would have paid off handsomely, yielding a profit of £66 when they won the 2002 Super Bowl. But our real point is that the probabilities in the table add up to almost 1.5, which at first glance makes no sense. It implies that there is a 150 percent chance that one of the teams will win the Super Bowl, way beyond metaphysical certitude. What's going on here? We can find out by practicing another of our habits and building a model.

Imagine that at the same time Ladbrokes had also offered the following odds that the Republicans and Democrats would control the U.S. Senate after the 2002 election. (Remember, a draw isn't possible, since the vice president gets to break ties.) Again, we've converted the odds into probabilities.

Party in Control	Odds	Probability
Republicans	1–3	0.75
Democrats	1–3	0.75

Here, too, the probabilities add up to 1.5. That's just as non-sensical as before, as both parties can't have a ¾ chance of controlling the Senate, no matter what Jim Jeffords does. But this simpler model makes the bookie's arithmetic much more transparent. Suppose someone bets £3 on the Democrats, while another bettor stakes £3 on the Republicans. Whichever side wins, the other comes away empty-handed, so Ladbrokes receives £6 but only has to pay out £4 (the winner's wager plus the £1 profit mandated by the 1–3 odds). Bully for Ladbrokes, which pockets one-third of the sum wagered. Hard cheese for the gamblers, who can expect to lose 33 pence on every pound they bet.

The Arithmetic of Probabilities

Many probability calculations are composite in nature. Unfortunately, the rules behind those calculations tend to get cloaked in unfriendly jargon. According to the vernacular of probability courses, the probabilities of disjoint events should get added, while the probabilities of independent events should get multiplied. But what does all that mean?

As is so often the case with mathematical terminology, the bark of probability jargon is far worse than its bite. For example, "disjoint" events are simple to describe. If you choose a single number in a game of roulette, your chances of winning are small: just $\frac{1}{38}$ on an American wheel, as we saw earlier. If another bettor chooses a different number, the two numbers are considered "disjoint events" and the probability that one of the gamblers will win is simply the sum of the probabilities of the individual events. Thus, the probability of the ball landing on the number 5 *or* the number 27 is $\frac{1}{38} + \frac{1}{38} = \frac{1}{19}$. That's just what our intuition would tell us.

Independent events are different. "Independent" means unrelated. To wit, the chance of a coin flip coming out "heads" is $\frac{1}{2}$. The chance of plucking the jack of diamonds from a deck of cards is $\frac{1}{52}$. The likelihood of getting a "heads" *and* picking the jack of diamonds equals $(\frac{1}{2})(\frac{1}{52}) = \frac{1}{104}$. Probabilities of independent events must be multiplied. So far, so good. (Not that you'd ever be picking cards and coins simultaneously, but probability courses are filled with props such as coins and cards, perhaps because the entire subject was born in an effort to compute gambling odds.)

Where our intuition fails us is when assessing the probabilities of *disjunctive* events, which are actually familiar and ordinary despite their obscure name. Suppose that a factory machine contains 10 key parts, and that for any particular shift, the probability that any given part performs perfectly is 90 percent. What are the chances that one or more parts will fail during a shift, causing the machine to break down?

The good news is that these parts aren't like slots on a roulette wheel. If they were, the chance of failure would be 10 times 10 percent, or 100 percent, and that's obviously impossible. We need a different approach, based on the simple insight that for the machine to operate perfectly, every single part must operate perfectly. Because the parts are assumed to be independent, the chance of all of them operating perfectly equals $(\frac{9}{10})^{10} = 0.349$. The chance of one or more parts breaking down is therefore $1 - 0.349$, or 0.651: a hefty 65.1 percent. In short, 90 percent reliability for a machine part is pretty miserable.

By themselves, probabilities may not give you much perspective on their magnitude. If a casino game offered you a 1 in 38 chance of winning, as roulette does when you bet on a specific number, would that probability be big or small, favorable or unfavorable? It all depends on the size of the potential payoff relative to the wager. If a successful $1 wager paid $1,000, then a $\frac{1}{38}$ chance of winning is a big number, whereas if the payoff were $10, the same probability would be small. But how big and how small? To get that fine a perspective, you need another probabilistic construct, called *expected value*, which in essence represents the average outcome. In roulette, the actual net payoff on a winning $1 single-number bet is $35. So when making such a bet you have a $\frac{37}{38}$ chance of losing $1 and a $\frac{1}{38}$ chance of winning $35. The expected value, obtained by weighting the different possible outcomes by their respective probabilities, is $(\frac{37}{38})(-\$1)$ + $(\frac{1}{38})(\$35)$ = -$0.053. What this means is that you will lose an average of 5.3 cents on each $1 bet. Thus, if a roulette player bets $1,000 over the course of an evening and only loses $20, he's had a pretty good night, because his expected loss was $53.

You don't have to go all the way to a casino to find numerical applications for expected value. In fact, you don't even have to leave home. Glance into your pantry and you might find a cereal box that contains one of several possible prizes as part of a promotional giveaway. Let's say there are six different prizes. The question is, on average, how many boxes of cereal would you have to buy before you had "won" all six prizes?

This problem is by now classic and has a nice, compact solution. The derivation is beyond the scope of this book, so you'll have to take our word that the expected number of boxes is not 6; it is $6(1 + \frac{1}{2} + \frac{1}{3} + \frac{1}{4} + \frac{1}{5} + \frac{1}{6})$ = 14.7. Part of the effectiveness of such promotions lies with the fact that peoples' estimates of these expected values are lower than the actual number. And as you add more and more prizes, this divergence only increases. In fact, one of the first paradoxes learned by calculus students is that when you add the fractions $1 + \frac{1}{2} + \frac{1}{3} + \frac{1}{4} + \frac{1}{5} + \ldots$, the result is unbounded, meaning that the sum will eventually top one billion, one trillion, or any other big number you could mention.

Promotions don't delve into such huge numbers, but we can say that a 2002 Topps Total Baseball set consisted of 990 cards. Assuming that the individual cards are in equal supply, you'd need to collect an average of 7,401 cards to get a complete set. No wonder card trading became so popular.

Although it's unlikely that you'll have many occasions to formally compute an expected value, a general understanding of the concept is nonetheless essential. Expected value serves as a reminder to consider not only the probabilities of different outcomes, but also their respective costs and benefits. Earlier in this chapter we mentioned a letter written by the director of the Aircraft Owners and Pilots Association defending the safety of general aviation (GA). Included in the letter was the following probabilistic comparison: "On a per-mile basis, you are seven times more likely to be involved in an automobile accident than a general aviation accident." That statistic may be accurate, but it doesn't indicate much about the relative risks of GA and automobile travel. Remember the moral of expected value: We can't assess the size of certain probabilities unless we know the payoffs associated with them. What that means here is that we need information about how injurious GA accidents are compared to automobile accidents. And when we look at that information, the statistic that car crashes are seven times as likely as GA crashes quickly loses force.

In 2000 there were 1,654 GA accidents, of which 342 were fatal.[13] In other words, 1 out of every 4.8 accidents was deadly. In the same year, there were 6.4 million motor vehicle accidents reported to the police, including 37,409 fatal accidents.[14] That's about 1 in 170. So GA accidents are around 35 times more likely to be fatal than motor vehicle accidents. But since GA crashes are only 1/7th as common per mile as automobile crashes, we reach the conclusion that flying in a small plane is five times as deadly as driving on a per-mile basis.

Second-Guessing Yourself

We met at nine.
(It was eight.)
I was on time.
(No, you were late.)
Ah yes, I remember it well.

We dined with friends.
(We ate alone.)
A tenor sang.
(A baritone.)
Ah yes, I remember it well.
                              —Gigi

In New York City, many car owners spend hours every week searching and waiting for a parking space. Other drivers avoid this hassle by paying garages large sums to scratch and dent their vehicles in the name of storage. These alternatives prompt millions of New Yorkers, one of us included, to use buses, subways, and taxis. But not your author's wife. She cheerfully drives herself around the Big Apple, secure in the knowledge that when it comes to finding a parking space, she is "lucky."

Your author is not so fortunate. He often has difficulty finding a parking space, and when he travels by air, his flights invariably depart from and arrive at whatever gate is furthest from the main terminal. And his connections couldn't be less convenient: Arriving at Gate A18 and then departing from Gate E33 at Hartsfield Atlanta would be a typical arrangement.

We have many friends and relatives who have a remarkable knack for picking stocks. They are not wealthy, mind you, but that's only because they didn't act on their best selections. "I *knew* back in 1986 that Microsoft was going to be a home run," they say. "I just never got around to actually buying the stock." A minor technicality.

Can you say "selective memory"? We knew you could. We all suffer from selective memory, and it leads us to believe silly things about the likelihood of certain events. Your author's wife isn't a lucky parker; she just remembers swift parking experiences and forgets the nightmares. Your author isn't an ill-fated air traveler; he selectively recalls flights at inconvenient gates. And our friends and family aren't brilliant stockpickers. They remember it well when they considered buying America Online in 1994, while conveniently suppressing all recollection of their purchases of Kmart and Pets.com, about which they were every bit as confident.

Selective memory is only one of many cognitive biases that make it hard for us to be good oddsmakers. Psychologists have documented numerous mental errors that people systematically make when judging the likelihood of events. Good oddsmakers are aware that the mind plays such probabilistic tricks and they have learned to second-guess themselves accordingly. It is worth noting a few of the more common cognitive biases so that you can get a flavor of the kinds of errors to which most of us are predisposed.

For instance, individuals tend to overestimate the probability of events whose past occurrence is easily or vividly recalled. Between 1990 and 2000, only five people were killed by sharks in the United States.[15] But because shark attacks are highly publicized and conjure up images from *Jaws* (together with John Williams's score), many beachgoers are terrified of swimming in the ocean. Yet these same people cross the street without a care, even though some 5,000 pedestrians are killed by motor vehicles in the United States every year.

People also tend to overestimate the likelihood of events that readily match stereotypes or other conventional images. In one often-cited study, subjects were given a description of a woman named Linda, who among other things was "deeply concerned with issues of discrimination and social justice, and also participated in anti-nuclear demonstrations." When asked to rank the probability of certain statements about Linda, subjects judged it more likely that "Linda is a bank teller and is active in the feminist movement" than "Linda is a bank teller."[16] That's illogical—the first statement can't be more likely than the second—but feminist activity fit Linda's profile better than tellership.

And did we mention that we humans are overconfident? In one experiment, subjects were asked simple factual questions, such as "Is Quito the capital of Ecuador," and then asked to estimate the probability that their answer was correct.[17] Subjects were overconfident; even those who were 100 percent certain of their replies were wrong about a fifth of the time. Men, by the way, tend to be more overconfident than women and at least in some contexts pay a higher price for their excessive self-assurance. Reviewing over 35,000 brokerage accounts, finance professors Brad Barber and Terrance Odean discovered that men traded stocks more frequently than women. No surprise there—if men are more likely than women to believe they can outsmart other investors, we would expect men to buy and sell more often. But the men were miserable stock pickers—the stocks they sold earned higher subsequent returns than the stocks they bought—so the more they traded, the worse they did. As a result, women's investment returns outpaced men's by nearly a percentage point per year.[18] That's not to say the women were such winners. The stocks they sold also earned higher returns than the stocks they bought.

Bayes Watch

Years ago a study asked the following question of students and doctors at Harvard Medical School:

> If a test to detect a disease whose prevalence is 1/1000 has a false positive rate of 5 percent, what is the chance that a person found to have a positive result actually has the disease, assuming that you know nothing about the person's symptoms or signs?[19]

The most common answer was 95 percent. The correct answer is 2 percent, and it was given by only 18 percent of the alleged medical experts. It's plausible that a number of those queried forgot that the "false positive rate" is the proportion of disease-free patients who test positive, and not the proportion of positive results that are incorrect, a misconception that could account for some of the wrong answers. (Not that such ignorance of basic medical terminology would be reassuring.) Nevertheless, it's likely that most of the Harvard group made the common mistake of neglecting the "base rate," the prior probability of an event. Here, the base rate is that only one in a thousand people has the disease. But if only 0.1 percent of the population has the disease, while 5 percent of the rest will test positive for it, then it should be clear that all else being equal, the overwhelming majority of positive tests will involve healthy individuals.

There's a major postscript to the Harvard Medical School study. Psychologists Leda Cosmides and John Tooby created a remake of the research, leaving probabilities unchanged from the original version, but otherwise substantially rewriting the story.[20] Subjects were asked the following:

1 out of every 1000 Americans has disease X. A test has
been developed to detect when a person has disease X.
Every time the test is given to a person who has the disease, the test comes out positive (i.e., the "true positive"
rate is 100%). But sometimes the test also comes out positive when it is given to a person who is completely
healthy. Specifically, out of every 1000 people who are
perfectly healthy, 50 of them test positive for the disease
(i.e., the "false positive" rate is 5%).

Imagine that we have assembled a random sample of
1000 Americans. They were selected by a lottery. Those
who conducted the lottery had no information about the
health status of any of these people.

How many people who test positive for the disease
will actually have the disease?

This time around, 56 percent of the subjects gave the correct answer of 2 percent. To Cosmides and Tooby, the dramatically improved performance demonstrates that people are actually good at
probabilistic reasoning when information is presented to them in
frequentist terms (say, "50 out of 1,000") rather than as single-event
probabilities (5 percent or 0.05), a view shared by other prominent
psychologists, including Steven Pinker and Gerd Gigerenzer.[21] And
it makes sense, according to Cosmides and Tooby, who argue that
our brains evolved to deal with probabilistic information as it was
observed in evolutionary times. "What was available in the environment in which we evolved was the encountered frequencies of actual
events—for example, that we were successful 5 out of the last 20
times that we hunted in the north canyon." Percentages, fractions,
and decimals were not to be found.

Frankly, we're a bit skeptical of the claim that humans are decent intuitive oddsmakers when dealing with frequencies. If the

people we know had succeeded on 5 of their last 20 hunting trips to the north canyon, most of them would "remember" having succeeded on 5 of their last 10 outings, conveniently forgetting two-thirds of their failures. Seeing frequencies through rose-colored glasses may be useful in some situations—perhaps it prevented hunter-gatherers from getting discouraged when food was scarce—but it doesn't make people good oddsmakers. And at least where the stock market is concerned, excessive hunting can be very costly.

As for 56 percent of subjects getting the right answer in the Cosmides and Tooby study, well of course people do much better on a problem when you walk them through the method of solving it, which is just what Cosmides and Tooby did. Calculating the exact answer to the question—$^{20}/_{1019}$—requires a somewhat cumbersome formula called Bayes' theorem, but you can do just as well with the back-of-the-envelope approach Cosmides and Tooby outlined. If 1,000 patients were tested for the disease, you would expect that 1 would have it and 999 wouldn't. The person with the disease would test positive, as would 5 percent of the 999, or roughly another 50. Thus, about 1 out of 51 people (1.96 percent) who test positive will actually have the disease. Pretty simple, isn't it?

And that's the main lesson we draw from the Cosmides and Tooby study. Regardless of whether we humans are superb, mediocre, or lousy intuitive statisticians, we all have an easier time with probabilities when we consider them as frequencies over a large number of trials, such as the testing of 1,000 patients. So you should get into the habit of converting single-event probabilities into frequency formats, especially when you're dealing with the interaction of multiple probabilities. Here's an example on which we can practice.

When O. J. Simpson stood trial for the 1994 murders of his ex-wife Nicole and Ronald Goldman, the country was treated not only to a yearlong melodrama, but also to one of the most shameless contortions of probability ever witnessed. Recall that one of the goals of the defense was to minimize the fact that Simpson had abused and threatened his wife during the years before her murder. Unfortunately for them, the admission into evidence of Nicole's harrowing 911 call of October 25, 1993, made their client look like the abusive, potentially murderous fellow that he was. Against that backdrop, defense attorney Alan Dershowitz wrote in the *Los Angeles Times* in January 1995 that although Simpson may indeed have abused his wife, the probability that an abuser goes on to be a murderer is about $\frac{1}{10}$ of 1 percent.[22] That's right: Even within the sordid group of husbands who abuse their wives, only one in a thousand actually goes on to kill his wife.

On the face of it, there was nothing wrong with Dershowitz's statistics. He noted that while roughly 1,500 women are murdered by a current or former mate every year, there are some 2–4 million spousal assaults. From these numbers, he surmised that only 1 in 1,000 abused women are killed annually. But in the spirit of playing *Jeopardy*, that's the right answer to the wrong question. Yes, had Professor Dershowitz been counseling Nicole Simpson in, say, November 1993, he could have said something like, "I know you're living in fear of your life, but statistically you're in good shape, because in any given year only one in a thousand abusers becomes a killer." But by January 1995 she was dead, and the only probability issue that mattered was the following one: Given that a woman who had been abused ends up murdered, what is the probability that the abuser turns out to be the murderer? Good-bye, one in a thousand.

To see exactly what's going on, it helps to think in terms of frequencies. And we don't even need to make our own model here, because Dershowitz's comments lit up a gigantic switchboard throughout the academic world. Jon Merz and Jonathan Caulkins began by observing that of the 4,936 women murdered in 1992, about 1,430 were killed by a present or past husband or boyfriend.[23] That's 29 percent, for those keeping score at home, even without any additional evidence. But Merz and Caulkins went further. Dershowitz had acknowledged that "among the small number of men who do kill their present or former mates, a considerable number did first assault them." Merz and Caulkins supposed that a "considerable number" meant half. They also assumed that 5 percent of women killed by someone other than a current or former lover were abused, a common estimate of the proportion for all women.

Where do we go with these various numbers? Merz and Caulkins applied a version of the aforementioned Bayes' theorem, but no such formula is required. Dr. Kevin Hayes of the University of Limerick showed the way with a simple table categorizing the 4,936 women who were murdered in 1992 according to whether they had been abused and whether they were murdered by a current or former husband or boyfriend:[24]

History	Murderer		
	Current or Former Husband/ Boyfriend	Other	Total
Abuse	715	175	890
No Abuse	715	3,330	4,045
Total	1,430	3,506	4,936

Before we draw any conclusions from this table, let's make sure we understand how Hayes got these numbers. The first column of figures came from the assumption that half of husband and boyfriend murderers were also abusers. The second column was derived from the assumption that 5 percent of the 3,506 women murdered by a non-boyfriend/non-husband had suffered abuse. Now, for the real eye-opener, turn your attention 90 degrees and look at the top row of numbers. There were an estimated 890 murders of abused women. Of those, 715 were committed by the abuser. On that basis, the probability that Nicole Simpson was killed by her abusive ex-husband Orenthal James Simpson was 715/890, or 80 percent. And that's even before Ron Goldman's blood showed up in his car.

Whether Dershowitz was mathematically ignorant or mathematically deceitful is not for us to say. Presumably the Dershowitzes and Cochrans and Schecks and Neufelds of the world maintain a clear conscience despite the outright fraud of the Simpson trial, because their duty is to defend their clients, not to uncover the truth. But society as a whole is still left with the responsibility of educating itself against other peoples' probability fraud.

Babylonian Oddsmaking

As we explained a couple of chapters back, good quantitative thinkers are Babylonians. They understand that quantities can be measured and expressed in various ways and that looking at something from multiple viewpoints enhances their perspective.

Babylonian thinking is especially important with probabilities, which often cloud reasoning even more than other forms of quantitative information. And there's good evidence that when the same probabilistic information is expressed in different ways, people sometimes reach quite different conclusions about its significance. For example, in a Swiss medical study, two groups of primary care physicians were shown findings from the Helsinki heart study on the effectiveness of cholesterol-lowering drugs.[25]

Both groups of doctors received information on the extent to which drug therapy reduced the risk of fatal and nonfatal heart attacks and death from all causes. However, one group saw this information presented in terms of percentage reductions, and the other group saw absolute numbers. For instance, where one group of physicians read, "A cholesterol lowering drug treatment reduces the relative risk of a fatal and non-fatal myocardial infarction by 34%," the other read, "A cholesterol lowering drug treatment reduces the incidence of fatal and non-fatal myocardial infarction by 14 per 1000 patients and five years of treatment."

The doctors were then asked to assess the effectiveness of lipid-lowering drugs and the likelihood that they would use them to treat a healthy male patient with total cholesterol of 288 mg/dl. Those who received the relative risk information regarded the drugs as significantly more effective, and were far more likely to prescribe the drugs than those who received the absolute risk information.

Remember, the point is not that medical risks are better presented in relative or absolute terms. These are two different perspectives, and doctors should consider both of them. The general lesson here is that you should try to look at all probabilistic information from different vantage points, especially when probabilities are presented as single-event chances. In 2000 there were 15.23 traffic fatalities per 100,000 U.S. residents, 21.94 fatalities per 100,000 licensed drivers, 19.27 fatalities per 100,000 registered vehicles, and 1.5 fatalities per 100,000 vehicle miles traveled.[26] All of these frequencies report the likelihood of dying in a traffic accident in 2000, but they provide different perspectives on that risk and together provide much more information about motor vehicle safety than any one of the statistics would.

Probability Distributions

Military historian D. M. Giangreco notes that when the military was developing casualty projections for potential U.S. invasions of Japan at the end of World War II, Army chief of staff George C. Marshall was reluctant to provide definitive estimates, and favored giving President Truman a range in which casualties might fall.[27] Good for Marshall, who presumably felt that single-point forecasts would mask the huge uncertainty about the outcome of invading the Japanese home islands and perhaps give Truman unwarranted confidence in the estimates.

Of course, where vicissitudes are concerned, the buck stops in the Oval Office, and Truman could hardly have presented his case to the American public in probabilistic terms. The public's need for clarity even applies to such everyday events as a weather forecast. Meteorologist Peter Leavitt, founder of the private forecasting firm Weather Services Corporation, laments that the forecast of a professional meteorologist contains much more equivocation and room for possibilities than an on-air presentation by a TV talking head. Society admires and demands decisiveness, even at a cost. We shout "Kill the umpire!" when we see a bad call, but the baseball umpire who takes a few extra seconds before making the correct call doesn't fare much better.

We have to get beyond the seductiveness of single-point estimates and recognize that all estimates of quantities are probabilistic, in the sense that there is a certain chance they will prove too high, or too low, or close to the mark, or way off. Standard household lightbulbs usually have a projected life of 750 hours. Yet we've all bought bulbs that burned out after only a few hours as well as bulbs that survived years of nearly constant use. To be sure, a basic bulb is more likely to last 750 hours than 1,000 hours, but 1,000 hours is not a surprising outcome. The point is that until the future plays out, the life span of a lightbulb, or the number of American lives that would have been lost invading Japan, or the number of minutes it will take you to get to work next Monday are not single numbers. They are distributions of possible outcomes, each with its own variations and contours.

In order to make wise decisions under conditions of uncertainty, it's essential to know as much as possible about that uncertainty. Isn't it remarkable, then, that almost all forecasts and other estimates are single-point estimates, which by definition provide absolutely no information about the uncertainty involved? What's more, those solitary numbers often show up unaccompanied by any account of what it is they are estimating or predicting, which can be quite different from what you might assume.

For example, the federal government's recommended daily intake of calcium for adults up to age fifty is 1,000 mg. Isn't it reasonable to assume that 1,000 mg represents an estimate of the daily calcium need of an average adult? It's not so. Nutritional balance studies show that an average adult can absorb only around 525 mg of calcium daily. But there's wide variation in needs, especially among women, who after menopause can often absorb more than 1,500 mg a day. So what question does the 1,000 mg figure answer? Well, here's how Recommended Dietary Allowance (RDA) is defined: "The average daily dietary nutrient intake level sufficient to meet the nutrient requirement of nearly all (97 to 98 percent) healthy individuals in a particular life stage and gender group."[28] In other words, the guideline is set so that if all adults age fifty or younger ingested 1,000 mg of calcium daily, 97 to 98 percent of them would be meeting or exceeding their needs.

Because single-point estimates conceal their own imprecision, they need to be followed up with additional research and thinking. Specifically, whenever you encounter a single-point estimate of a quantity, you should immediately take the following steps:

1. *Determine what the estimate is estimating.* If an information technology firm claims it will cost $2 million to upgrade your company's computer systems, is $2 million an estimate of the most likely cost? Or is $2 million the midpoint of a range of potential price tags? Or does $2 million represent an estimate of the upgrade cost if everything goes according to plan? Or is $2 million just a number designed to win your business?

2. *Approximate the probability distribution.* Remember, a single-point estimate is just a point on a probability distribution. While that point can convey valuable information, what you really want to know is the probabilities of the various possible outcomes. Play 20 Questions to that end. What's the chance that the project will run over budget? Under what circumstances would the project cost $3 million? What's the likelihood of that? How about $4 million?

3. *Assess your preferences.* It's better to arrive 15 minutes early for a movie than to miss the opening scene. Therefore, if it takes an average of 20 minutes to drive to the theater and get tickets, popcorn, and seats, you should leave for the theater earlier than 20 minutes prior to the movie time. How much earlier depends not only on the distribution of possible travel times, but also on how you feel about arriving earlier or later than you'd like to. If you're going to a movie you've been dying to see, you might insist on a 98 percent chance of getting to the theater on time; for a film you've already seen, perhaps you'd accept a 25 percent risk of being late.

Because single-point estimates don't allow for such calculations, people often forget to weigh their own preferences and establish an appropriate margin for error. Suppose a financial planner calculates that you need to save $16,000 a year to support your current lifestyle in retirement. Not only does that calculation paint over tremendous uncertainty in investment returns, longevity, financial needs, inflation, Social Security and Medicare benefits, and other economic circumstances, it also obscures the fact that it's better to save too much for retirement than to save too little.

4. *Seek superior information.* Trading a stock on inside information is lucrative because better information reduces the uncertainty about future share prices. The principle applies to any situation where there are unknowns—the better your information, the lower the uncertainty and the better your decisions. So bear in mind that every estimate can be improved by obtaining superior information. Just keep things kosher.

The Proof Is in the Numbers

Primary Colors, the thinly veiled fictional account of Bill Clinton's rise to the presidency, inspired a national guessing game upon its 1996 release. "Who is Anonymous?" everyone wanted to know. But the speculation all but ended when Vassar College English professor Donald Foster identified political columnist Joe Klein as the man behind the pseudonym. What gave weight to Foster's charge wasn't any conventional journalistic dope. No, Foster had something more; he had data. At the behest of *New York* magazine, Foster had conducted a statistical analysis comparing the vocabulary and writing style of Anonymous with those of the principal suspected authors.[1] Mindful that *while* versus *whilst* had distinguished Hamilton from Madison within *The Federalist* papers,[2] Foster found that Anonymous's vocabulary (such as an affinity for adjectives ending in "y") and style (lots of dashes, colons, capital letters, and short sentences) were far more similar to Klein's than anyone else's. Klein continued his denials, but Foster's work inspired others to pierce the facade once and for all. Five months later a *Washington Post* reporter uncovered a *Primary Colors* galley with Klein's handwriting in the margin. Klein confessed.

If you watch professional basketball, you've probably noticed that players are streaky. When a player has made his last several shots, he's in a groove and thus more likely than average to sink his next attempt. By contrast, when a player has missed a few in a row, his rhythm is off, which portends badly for his next shot. At least that's what we thought we observed until someone actually produced some data on basketball shooting. When Thomas Gilovich and his colleagues looked at all the shots taken by the Philadelphia 76ers over the course of a season, the data showed that shooting isn't streaky at all.[3] In fact, Sixers' players were more successful after misses than after hits. They made, for example, 56 percent of shots taken after missing their previous three attempts, and only 46 percent after making three in a row.

The unmasking of Joe Klein and the debunking of streak shooting are not simply amusing events. They are also episodes in a momentous story that we described at the outset of this book. In everything from literature to sports to finance to you-name-it, our knowledge and our decisions are increasingly based on the collection and interpretation of numerical data. What makes these two episodes noteworthy is that the raw data and the final conclusions are separated by a hidden set of judgments. That glossed-over, in-between step is the focal point for this chapter. It's called statistics.

Perhaps we should state the obvious: Not everyone likes statistics. And it's easy to see why. Statistical studies begin with piles of data, and data, apart from their inherent lack of charm, have the potential to be unclear, misleading, inaccurate, and sometimes downright fraudulent. (See Enron's financial statements or the federal budget.) Even when the data are untainted, statistical studies can seem heavy on inference and light on proof, to the point where those outside the field can feel hoodwinked. Famed nineteenth-century French physician Claude Bernard likened statistical reasoning to trying to understand what was happening inside a house by observing "how many people went into a house" and "how much smoke came out of the chimney."[4]

But think about the changes in the field of medicine since Bernard's time. In his day, and even well into the twentieth century, understanding the individual patient was the be-all and end-all of a doctor's world. Clinical practice and research were guided by physiological thinking about how diseases caused functional changes within the body. The profession couldn't possibly embrace statistical reasoning, which bypassed mechanistic understanding.

As medical science progressed, though, doctors began to appreciate that physiology is too complex to provide all the answers clinicians need. No amount of peering through a microscope is going to reveal for which types of patients and in what dosages the benefits of aspirin therapy (reduced risk of heart attacks, ischemic strokes, and possibly colon cancer) outweigh the adverse effects (increased risk of hemorrhagic strokes and major gastrointestinal bleeding). But if you gave different doses of aspirin to thousands of individuals and a placebo to thousands of others, you could find out. "The Second World War may be regarded as the great divide," wrote Oxford professor of clinical medicine Leslie John Witts in 1959, "after which it was no longer possible for the clinician, however distinguished, to discuss the prognosis and treatment of disease unless his words were supported by figures."[5] Today, in the Quantitative Information Age, it seems that almost everything learned about the effectiveness of medical treatments is the product of statistical studies.

If Claude Bernard's descendants can overcome their fear of statistics, so can we. For our purposes, "statistics" means simply the art of extracting information from a bunch of numbers. What makes someone a good quantitative thinker is the ability to extract this information—not just occasionally, but routinely. To make sure we keep things routine, we won't be delving into chi-square tests, box plots, Poisson distributions, or any of the other standard tools from a high school statistics course. With the

Quantitative Information Age as our backdrop, the central theme of this chapter can be summed up rather more simply. Repeat after us: "Numbers are evidence." Our humble goal is to reach conclusions and make decisions on the basis of the evidence we come across in our everyday lives.

The Power of Randomness

A crucial first step is to develop a healthy respect for the power of randomness. The notion that many events have no assignable cause is surprisingly difficult to accept. Perhaps that's because randomness gets confused with disorder and meaninglessness, both of which are discomforting to most of us. Our brains would rather create nonexistent patterns than be without patterns altogether. But the fact remains that much of what happens in the world is the product of chance. If we don't learn to appreciate the enormous influence of chance, we will regularly read too much into events and in so doing leap to conclusions that aren't justified.

To demonstrate what we're all up against, we created an imaginary baseball team and followed the batting averages of nine players: Ascione, Breuer, Costa, Deskin, Edwards, Frimerman, Gonzalez, Handlowitch, and Imparato. (The names were selected blindly from a phone book, with one name drawn from each of the first nine letters of the alphabet.) We started with the assumption that Ascione and the others have identical batting skills. They're all .275 hitters; that is, for each and every at-bat, there is a probability of 0.275 that the player will get a hit. From there, we had a computer simulate at-bats for the players by generating random numbers, which is what fingerless computers do in lieu of coin tossing.

Now, let's look at how the players fared in their rookie season. The table below summarizes each player's performance.

FIRST-SEASON AVERAGES

Player	At-Bats	Hits	Batting Average
Ascione	600	180	.300
Breuer	600	139	.232
Costa	600	156	.260
Deskin	600	171	.285
Edwards	600	176	.293
Frimerman	600	191	.318
Gonzalez	600	170	.283
Handlowitch	600	163	.272
Imparato	600	176	.293

Notice the dramatic effect of randomness. Breuer, for instance, hit 43 points below his true average of .275, while Frimerman hit 43 points better than expected. You can bet that few people looking at such figures in real life would consider the role that chance might have played. Instead, almost everyone would conclude that Breuer doesn't have what it takes and that Frimerman is a talented hitter with a bright future ahead of him. Breuer would get cut from the team and Frimerman would demand a hefty raise. All because Frimerman was lucky and Breuer unlucky.

We're not suggesting that batting averages reveal nothing about hitting skill. Tony Gwynn didn't win eight National League batting titles just by chance. Our point here is that batting averages, like many numerical outcomes, can be heavily influenced by randomness. As a witness to these numbers, you need to recognize this influence and discount some of the inferences you might otherwise draw. If you don't, you'll make unfortunate decisions, such as cutting Breuer, who our computer says went on to hit .302 the next season, the highest among the nine players.

Much of the fine art called statistics is actually an effort to identify, quantify, and deal with the random component found in any given set of numbers. The fact that randomness can undercut inferences and conclusions is certainly annoying, but we don't have to give up.

Back to Normal

Paradoxically, chance, the unpredictable element in events, can sometimes help you predict the future. Suppose a large group of high school students takes the SAT (by which we mean the exam now called the SAT I), and then, before any of them has a chance to study further, all take the SAT again. Not knowing anything about these students, we confidently predict that those with low scores on the first test will, more often than not, improve on the second test, while those with high initial scores will collectively do worse the second time around.

How can we make such an assertion? Think of it this way. There are two types of students who have low scores on the first exam: those whose scores are representative of their aptitude, and those whose scores are uncharacteristically low and who scored poorly due to bad luck. (Yes, there will be a cretinous few who scored 300 and were lucky to do so, but the shape of the Bell curve assures that their number is limited.) Luck is not a trivial factor: If .275 hitters like Breuer sometimes bat .232 or worse over 600 at-bats, then test takers will often score well below their true aptitudes over 138 multiple-choice questions. However, on average, luck is a neutral factor, and so looking ahead we have no reason to expect a given student to do better or worse than his aptitude merits. On the second test, then, those who scored appropriately low on the first test will tend to score poorly again, while those who initially under-performed are likely to improve, thereby raising the average performance of the entire group that had low first-exam scores. Similar reasoning applies to the students who scored well on the first SAT. Some deserved to do well; others were fortunate. The genuine high-scorers are apt to perform similarly on the second exam, while the lucky high-scorers will tend to do worse, dragging down the average score of the first-exam high-scorers.[6]

This phenomenon is typically called "regression to the mean," and its recognition is credited to English scientist Sir Francis Galton, who in growing sweet peas noticed that the progeny of heavier seeds were usually lighter than their parents, while the offspring of lighter seeds tended to be heavier than their parents.[7] The expressions "reversion to the mean" and "mean reversion" are also used to describe this effect. (Galton first used "reversion" and later "regression," but with the exception of a few statisticians, nobody draws distinctions among the various terms.)[8]

Going back to the SAT for a moment, regression to the mean is a key issue in the intense debate about the effect of formal tutoring on students' scores. The College Board has more than a little invested in the premise that the SAT is a fair and valid test that isn't biased in favor of those who can afford professional coaching, whereas companies such as Kaplan and Princeton Review depend on people believing that their courses, books, and software really work. Nobody disputes the fact that students who retake the SAT after coaching usually score better on their second try. But students who aren't coached also improve, albeit by less. This raises the possibility that gains among coached students are mainly the result of increased familiarity with the test, the greater knowledge that comes with age and experience, and regression to the mean.[9] Presumably students who scored poorly relative to their true abilities are more likely to retake the SAT than students who did unexpectedly well, and it wouldn't be surprising if the biggest underperformers were especially likely to seek coaching. If that's the case, then increased familiarity, aging, and regression to the mean could produce considerable improvement among coached students, even if the coaching were completely ineffective.

Unhappily for parents of high school students, existing research can't really tell us how much coaching adds to performance
on top of these other factors, although it wouldn't be difficult to
design a study that did so. As Harvard epidemiologist James
Robins explained to the *New York Times Magazine*:

> All it would take to get credible results on the effects of
> coaching is a very simple study. You hire an independent
> researcher to find several hundred kids who weren't plan-
> ning to be coached. Give them the S.A.T.'s. Flip a coin.
> Those who come up heads get coaching of one kind or an-
> other. The rest don't get coaching. Then you see what
> happens. It's not a big deal. Maybe the real problem is
> that everyone is a little nervous about what the study
> would find.[10]

Once you start looking for regression to the mean, it shows up
all over the place. Consider the pattern known as the "SI Jinx,"
the apparent curse visited upon athletes who appear on the cover
of *Sports Illustrated*. Time and again, athletes who have been the
subjects of *Sports Illustrated* cover stories find that their performance subsequently tails off. (In researching a cover story on its
own jinx, *Sports Illustrated* found 913 such jinxes in 2,456 cov-
ers.)[11] Curious, isn't it? Not to students of regression to the mean.

Recall our hypothetical baseball team of nine rookies, all of
them .275 hitters. Say that halfway through the season, *Sports Illustrated* decides to do a cover story on the team's "best" rookie. In
our computer simulation, that would be Edwards, who was batting a torrid .327 at that point in his then-brief career. Predictably, though, Edwards performs closer to his true ability in
the second half of the season, batting .260 and providing believers with another confirmation of the jinx.

As Peter Bernstein notes in his best-selling book *Against the Gods*, regression to the mean "explains why pride goeth before a fall and why clouds tend to have silver linings. Whenever we make any decision based on the expectation that matters will return to 'normal,' we are employing the notion of regression to the mean."[12] Banking on a return to normal might sound like pretty basic stuff, but decisions are all too often based on precisely the opposite expectation. Buyers of mutual funds seem wedded to the streak-shooting concept. The fund manager who has the hot hand will find it especially easy to attract new investors. By contrast, a manager who has had three straight lackluster years can't very well persuade his firm's marketing department to proclaim "Buy Henderson's fund: He's due!" (Henderson might not have been doing anything wrong. Visualize a value fund manager in early 2000, just before the NASDAQ bubble burst.) As we saw in chapter 4, the *average* investor's experience with a fund can be negative even when that same fund's net move is positive, simply because more people buy in after the surges than after the setbacks.

But students of regression to the mean are not surprised when the "best" mutual fund managers become ordinary performers or, for that matter, when friends report successfully fighting off a terrible cold with truckloads of echinacea. At any given time the best performing money managers are also the luckiest ones, and if bad colds didn't normally get better, you wouldn't be around to read this.

Correlation and Causation

Smoking is strongly correlated with lung cancer. The more someone smokes, the higher the risk of developing lung cancer. The size of children's feet is strongly correlated with their reading skills.

On average, the bigger a child's feet, the better the child reads. The first correlation indicates a causal connection; the second correlation doesn't. Smoking is a cause of lung cancer. Bigger feet don't cause improved reading, or vice versa. Aging is the common cause of the physical and mental development that drives both shoe sizes and reading levels.

That the association between smoking and lung cancer is evidence of a causal link, whereas the association between shoe sizes and reading skill is not, well, that's pretty obvious, but only because we know the truth in advance. Normally when we look at associations among data, we don't know the facts underlying the numbers. A safety-minded car shopper browsing the Insurance Institute for Highway Safety (IIHS) website will learn that Volvos have lower injury and death rates than comparably sized Chevrolets. But to what extent does that mean Volvos are structurally safer cars than Chevrolets? No offense to Chevy drivers, but wouldn't you figure that Volvos, which are stereotypically owned by safety-conscious families, are driven more carefully than Chevys? And for all we know, Chevys average more miles annually than Volvos. Since the IIHS injury and death rates are calculated per vehicle year—and not per vehicle mile, as many traffic safety statistics are—couldn't differences in mileage be part of the story?

Asking these sorts of questions will help you make sense of associations in data, as will the process of trying to answer them. Realistically, though, that process will often amount to no more than making reasoned guesses. A car shopper who has the time to investigate and analyze differences between Volvo and Chevy drivers is probably unemployed and should get a job before considering the purchase of a car. Don't get us wrong—we're not trying to put a damper on your inquisitiveness; that's the last thing we would want to do. We just want you to recognize that educated guessing will sometimes be a more efficient use of your time than painstaking research.

Nevertheless, it's worth understanding a little bit about how statisticians, who get paid to do painstaking research, try to overcome the ambiguity inherent in basic associations. One of the tricks of the trade is to seek out data that are less equivocal. To assess the relative safety of Chevys and Volvos, a statistician might look for a risk measure less influenced by driving characteristics—deaths per accident, for example, instead of deaths per vehicle year.

A more advanced step is to take into account the influence of factors other than the one being investigated. To judge how much safer Volvos are than Chevys, a statistician would need to consider the impact of driver behavior on injury and fatality rates. To do this, a statistician might gather data that would (directly and indirectly) indicate the driving habits, driving skill, and safety consciousness of Volvo and Chevy owners, such as annual mileage, frequency of accidents, speeding tickets and DUI convictions, age, sex, education, marital status, parental status. In other words, the very data car insurance companies want to know when they set individual rates. With such data, a statistician could then use a mathematical technique known as multiple regression analysis, which essentially attempts to ask and answer the following question: If we compare Volvo and Chevy drivers with identical characteristics—same annual mileage, accident rate, speeding ticket frequency, DUI record, age, education level, sex, marital status, and parental status—to what extent do the Volvo drivers get injured and killed less often? The answer is an estimate of how much safer Volvos truly are.

Of course, regression analysis can't really make Volvo and Chevy drivers identical. What the technique does is estimate the degree to which differences in injury and fatality rates between Volvos and Chevys are attributable to the cars themselves as opposed to how they are driven. That kind of estimation is often quite effective, making regression analysis one of the most useful implements in a statistician's toolbox. But regression analysis is inadequate for many tasks, and frequently researchers go a step further and create experiments that actually try to make different groups—such as Volvo and Chevy drivers—exactly alike.

Way back in the second chapter we mentioned that in the late 1950s, Sir R. A. Fisher, who was perhaps the world's leading statistician, argued, at great cost to his reputation, that the evidence linking cigarette smoking to cancer was only circumstantial. Fisher's defenders claim that he was trying to make a methodological point: Cause and effect could not be proved by simple correlations. Fisher, who had been knighted for his work in genetics, pointed out that the observed correlation between smoking and lung cancer might be explained by a genetic tendency toward both. That is, if there is something genetic that makes some people more likely to smoke than others, and if that same hereditary factor also places such people at particular risk of developing lung cancer, then genes rather than smoking could be responsible for the high rates of lung cancer among smokers.

In Fisher's view, casual proof required a randomized experimental design, such as the one James Robins proposed to determine the effect of coaching on SAT scores. In its basic form, a randomized controlled experiment works like this. A sample of representative subjects—be they people, lab rats, plants, what have you—is randomly divided into two groups. One group, typically called the "treatment group," receives the intervention that is being tested, such as SAT coaching, a baby aspirin every day, or a new fertilizer. The other group, the "control group," receives no treatment. Where feasible, the control group receives a placebo, such as a sugar pill, so that its members don't know whether or not they are receiving the treatment.

The aim of a randomized controlled experiment is to make the two experimental groups comparable in every respect except that one group receives the tested intervention and the other group doesn't. That way, any differences in outcomes—SAT scores, heart attacks, plant growth—can be attributed to the treatment in question. Remember why we can't just compare the relative performance of those who do and don't get SAT coaching in real life: The two groups may be different in ways that could affect scores. But with a randomized controlled experiment, we hope to eliminate that possibility, since the determination of who gets coaching is made by a flipped coin.

As a historical note, it was Fisher who pioneered the use of randomization as a method for treatment assignment. However, Scottish physician James Lind is usually credited with conducting the first experiment in which subjects were systematically given different treatments. In 1747, Lind selected twelve sailors suffering from scurvy on a British warship and, in groups of two, put them on six different diets. "The consequence was," Lind later wrote in his *Treatise of the Scurvy*, "that the most sudden and visible good effects were perceived from the use of the oranges and lemons; one of those who had taken them, being at the end of six days fit for duty."[13] Unfortunately, the Admiralty was slow to recognize the importance of Lind's finding, and it wasn't until 1796 that the Royal Navy began to provide sailors on long voyages with a daily ration of lemon juice. Thereafter, incidence of scurvy declined dramatically. Years later, lime juice was also used, giving rise to the term "limey."[14]

Although developed with scientific research in mind, randomized controlled experiments have become a mainstay of marketing research, most notably in direct marketing. For instance, companies that want to assess the relative effectiveness of different catalogs will often conduct experiments by sending different catalogs to different groups of randomly selected customers. Similarly, telemarketers regularly use randomized experiments to test alternative scripts or promotions.

Just knowing the principles of controlled experiments can help us unravel some paradoxes of correlation versus causality. For example, bad backs are often attributed to structural abnormalities such as herniated, bulging, or degenerative disks. When a patient with back pain sees an orthopedist, a common first step is an X-ray or MRI to identify such abnormalities. The next step might be surgery to fuse or eliminate the apparently offending disk. In many cases the symptoms dissipate. Cause and effect, right?

Sorry, but it's not that easy. For one thing, there's the now-familiar regression to the mean. Back problems have their ups and downs, and since people seek medical help when their pain is at its worst, their condition will often improve after seeing a doctor, with or without surgery. But can't we still assume that surgery accomplishes something by correcting an underlying abnormality? Nope: For it's not clear that the disk problems most back surgery aims to alleviate are necessarily hurtful in the first place. MRI scans reveal that people free of back pain have high rates of disk abnormalities, just as back pain sufferers do.[15] This evidence that abnormalities are normal bolsters the view of some physicians, such as Dr. John Sarno of NYU's Rusk Institute, who believe that back pain is usually the result of mental and emotional stress, not physical imperfections. And if that hypothesis is correct, then surgery will sometimes work not because of its physical effects, but because of its emotional impact; by convincing patients that their back problems have been solved, surgery relieves stress, much like a hot bath, massage, or glass of wine.

Back pain sufferers and everyone else should be grateful that Fisher invented and championed the randomized controlled experiment. Yet randomized controlled experiments can't solve everything, not only because they are expensive and time-consuming, but also because they're sometimes unethical or unfeasible. To investigate the link between smoking and cancer to Fisher's satisfaction would require recruiting a group of non-smokers and then randomly assigning half of them to decades of smoking. Or how about global warming? Can you think of a way to randomly change the CO_2 concentration in half of the world's atmosphere?

You get a gold star if you spotted another difficulty: How can we ever be sure that differences in outcomes between the treatment and control groups aren't the product of chance? Weeding out this possibility is the crucial next step.

Testing the Hypothesis

Let's say we were to conduct the SAT coaching study that Professor Robins put forward. We recruit a bunch of kids who weren't planning to be coached (which avoids the ethical problem of taking coaching away from someone), give them the SAT, split them by random into two groups, send one group to Kaplan or Princeton Review, and then retest both groups. For the sake of argument let's assume that the coached kids improve their scores by 20 points more, on average, than the uncoached kids. Does that prove that coaching works? Isn't it possible that the coached group was unlucky on the first test and lucky on the second, or that luck worked the other way around for the uncoached group? Isn't it also possible that, by chance, the coached group and uncoached groups turned out to be a little different? Maybe the coached group had more motivated kids who, even without coaching, would have studied somewhat harder than the others. Clearly we need numbers that can put these questions to rest, but where do they come from?

The source of those much-needed numbers is another statistical tool, commonly known as hypothesis testing. What we do is assume that coaching has no positive effect on scores. This assumption is called the null hypothesis. Then we ask: If coaching were ineffective, how likely is it that the group of coached kids would improve by 20 points more than the uncoached group, just by chance? Fortunately, this probability can be calculated using basic statistical formulae. (We said we wouldn't bring up chi-square tests. We lied. That's one of the techniques behind the calculations that statisticians make at this pivotal stage.) If the probability that chance alone would produce a 20-point or greater differential is very low, we take a deep breath, cross our fingers, and conclude that chance didn't cause all of the 20-point difference. In statistics jargon, we reject the null hypothesis and accept its alternative, that the SAT coaching is effective.

You get another gold star if you're wondering what we mean by a "very low" probability. A probability of 5 percent is the most common cutoff. In other words, if there is less than a 5 percent likelihood that an outcome such as the one we observed would occur by chance, then we assume that the outcome wasn't solely the product of chance. When you read in the newspaper that a finding in some study was "statistically significant," it usually means that a result of such magnitude would occur by chance less than 5 percent of the time. (The 5 percent benchmark dates back to an era without computers, when statistical calculations were far more difficult to make. Nowadays we can readily compute probabilities relative to any threshold level we choose, but the 5 percent level has withstood the test of time and progress.)

The logic of statistical testing may sound convoluted, but it's actually second nature for all of us. If a schoolchild says that his dog ate his homework, his teacher will not believe him. Why? Surely dogs occasionally eat homework. (Both of us have had dogs that ate far less appetizing and accessible items than homework— in one memorable case, the snack was a check for $3,000.) But the teacher asks herself: What's the likelihood that a given assignment gets consumed by a dog? Answer: Really, really low. So the teacher rejects the hypothesis that the kid is telling the truth in favor of the alternative, that the kid's lying. Next stop, detention.

Sample Sizes

As we described the procedures of the previous section, we omitted one basic, intuitive truth: The influence of randomness wanes
as the number of events increases. If we ever carried out that controlled experiment on SAT scores, we'd want to recruit as many
students as we could if we wanted to get below that magic 5 percent level of statistical significance.

To provide some actual numbers on the effects of larger sample
sizes, we first go back to our baseball example and the following
table:

LIKELIHOOD OF A .275 HITTER EXCEEDING EXPECTATIONS		
Number of at-bats	Probability of .280+ average	Probability of .300+ average
100	49%	32%
1,000	37%	4%
5,000	22%	0%

These probabilities are the very sort of "take someone else's
word for it" calculations that make statistics so frustratingly dis-
tant, but at least we can say that the numbers are based on well-
established formulas. (Data often fall into familiar distributions,
or patterns, the most familiar being the normal distribution of the
Bell curve.) More important than the specific numbers are the
patterns they produce. Moving downward within either probability column, we see that the deviation from a known average gets
progressively less likely as the number of at-bats goes up. Comparing the two columns, we see that for any fixed number of at-
bats, the likelihood of a small deviation is much larger than the
likelihood of a large deviation.

The foregoing doesn't exactly qualify as rocket science, but in
the absence of such a simple visual, the impact of sample sizes
is often forgotten. Some years ago a consulting firm conducted a

study for a bank on Long Island. The premise of the research, resulting from a series of preliminary questionnaires, was that the bank had an unusually high market share—25 percent—among the small to medium-sized businesses in one specific area. If only they could find out what they were doing right, perhaps the bank could raise its share throughout Long Island. Sound reasonable? Before answering that question, we should say that the preliminary results triggered a spirited campaign of telephone interviewing in which hundreds of small business owners were asked about their banking preferences. Alas, the only meaningful conclusion was that almost all of the businesses chose their bank because it was the closest. Even worse, the alleged 25 percent market share was a phantom; the bank's share within the enlarged sample was only 10 percent.

If you're suspecting that the preliminary questionnaires weren't geographically spread apart, you're right, but the problem was worse than that. It turned out that the alleged 25 percent share was derived from a set of 28 questionnaires. That's right. Of an original sample of 28 respondents, seven did their business with the client bank, but because the result was communicated to the marketing department as a percentage, the small sample size was forgotten and a major study ensued. The arithmetic that yielded the 25 percent figure was flawless, but the weight placed on such a small sample was absurd.

There is a troubling irony brewing here. It's easy to see that a sample of 28 questionnaires was ludicrously small, and it's fun to ridicule a marketing plan gone bad. Yet most of us draw on much smaller samples as an unconscious way of life. Just ask our friend who owns a Glen of Imaal terrier. The breed is uncommon, sort of in between a Wheaten terrier and a West Highland white. This particular representative of the species, Molly, happens to have an underbite that would keep her out of the show ring for life. And what do visitors to the house say when they get their first glimpse of a Glen of Imaal terrier? Nine times out of ten they ask something like, "Do all Glen of Imaal terriers have teeth like that?"

Replace terriers with humans and some of the humor leaves the story. Someone whose parents always fought with one another may grow up believing that parents are fundamentally nasty and argumentative, never mind the sample space of one. This built-in tendency to extrapolate causes even bigger problems when applied outside the family. If a man in your neighborhood happens to be, say, Filipino, and you see him running a red light, you will be considered politically incorrect if you conclude that Filipinos are reckless drivers. However, what makes the statement troublesome isn't the accusation itself: The alternative statement, "Old men wearing hats are poor drivers," makes a similar claim (in this case, we think, accurately), but by shifting the focus away from an ethnic group, the accusation now seems benign. No, the real problem with many statements about ethnic groups is that they rest on puny sample spaces. It's human nature to stretch observations about two or three people so as to fit an entire population, which is what stereotyping is all about. Note that the sample space issue exists for positive attributes as well as negative ones. You can say "Filipinos are generous people" based on your same neighbor giving your kids Snickers bars at Halloween, but the sample space problem is identical.

To eliminate unwarranted stereotyping, we must expand our knowledge of other societies, but even those efforts are hampered because of a potential overdependence on one specific reporter or writer. When we read accounts of life in certain foreign countries, our quest for fresh perspective can make us forget that we are getting just one person's view. On a different note, we've all seen statistics that compare the performance of the stock market during Republican administrations with that same performance during

Democratic administrations. But even if we accept the notion that the Dow Jones Industrial Average is firmly correlated with the party in the White House—a dubious proposition, because of the hundreds of other variables at work—we're still talking about only a few administrations on either side during the twentieth century. (For the record, the average annual gain in the S&P 500 with a Republican at the helm is only 8.0 percent, way behind the average 12.3 percent annual gain posted by the Democrats. These figures may be a surprise to those who link Republicans with big business, but they make sense when you consider that the 1929 crash, the 1970s oil shock, and, for that matter, the 1987 correction, all took place during Republican administrations. A few data points swing the entire result.)

Although the costs of undertaking surveys and experiments places an upper bound on all sorts of sample sizes, we finally have some good news. Even a relatively small sample of a large population, *provided that sample is truly random*, can approximate the larger population with uncanny accuracy.

Online polling, for all its flaws, gives us a chance to see this principle in action. America Online users are frequently polled for their opinions on current events and other issues. One of the questions posed in 2001 was, "Do cell phone users annoy you in public places, such as restaurants or movie theaters?"

You may be wondering who *isn't* annoyed by cell phones in movie theaters. Indeed. Polls are notorious for phrasing their questions in a manner that all but dictates the final outcome, and this particular outcome was hardly a surprise. After 1,545 responses, the responses showed a clear trend:

Yes	1,115	72.2%
No	154	10.0%
Not sure	276	17.9%

However, suppose you added another 5,782 respondents. Intuition suggests that the results might change quite a bit, because the new total of 7,327 is almost five times as big as the original sample of 1,545. However, unless there was something unusual about the first 1,545 respondents (and there is no reason to think that the early participants in the poll had different feelings about cell phones than the later ones), the results were all but set in stone. The precise figures at the 7,327 mark were remarkably similar to the first batch.

Yes	5,216	71.2%
No	745	10.2%
Not sure	1,366	18.6%

We stopped tracking at that point, confident that nothing would change, but you can see that we had already gone on longer than we had to. Random samples converge to their targets quite rapidly, a principle that goes by many names, from the "law of large numbers" to "central tendency." Whatever you call it, rapid convergence has long been one of the most important results in the entire field of statistics and probability. Even if 100 million people ended up taking the poll, the distribution of their responses would be unlikely to diverge significantly from a sample of 7,500.

Polls typically express their expected accuracy via a sampling error. The most common example is a political poll that shows Candidate A with a 47 percent to 42 percent advantage over Candidate B, with a sampling error of, say, plus or minus 3 percentage points. What does a sampling error of 3 percentage points mean in this context? Something like this: 95 percent of the time, the actual percentage of the population that favors a given candidate is within 3 points of the figure indicated in the poll. All in all, we'd rather be Candidate A, but in these cases the real question is not the size of the sample, but whether that sample is truly representative of those who will actually vote. The law of large numbers, plus or minus three points and all, applies only to random samples, which can be frustratingly difficult to achieve in real life.

Interestingly, the most sophisticated political polling doesn't involve purely random sampling. Let's suppose that half the likely voters in a given election are Democrats, and half are Republicans. A random sample of 500 voters might turn out to have 270 Republicans and 230 Democrats, which is unlikely to give an accurate prediction if preferences fall strongly along party lines. In that case, you would do better by randomly selecting 250 Democrats and 250 Republicans, or by weighting results to adjust for any bias in party membership. In practice, polling organizations such as Gallup do this, not just with party affiliation but also with race, gender, age, and other characteristics known to be strong predictors of voting behavior.

But if the percentages used to make such extrapolations aren't accurate, everything falls apart. One of the pivotal errors surrounding the Florida projections in the 2000 presidential election was an assumption that absentee ballots, which skew Republican, would constitute 7.2 percent of the state's total vote. The actual absentee representation in Florida was more like 12 percent, the net effect of which was a temporary 1.7 percentage-point error in Gore's favor. However, even if the data had been perfect, Florida would have been "too close to call" the entire night and even beyond. Polling and projections were never designed to pick a winner in an election that tight.

Although complaints about misleading polls are understandable, human factors are precisely what guarantee their inaccuracy. With many people replacing home phones with mobile phones (pollsters aren't allowed to call mobile phones), and others increasingly screening their calls with answering machines and caller ID, pollsters are having a harder and harder time sampling voters, and in turn obtaining a sample that pollsters believe is representative. And beyond the mathematics of it all, who among us is completely accurate in forecasting our future behavior? Polls are but a momentary still frame in a motion picture where only the last frame counts.

The final straw for pre-election polls is that they by definition neglect get-out-the-vote efforts, or GOTV. Although Republicans raise more money for campaigns than Democrats do, Democrats (thanks to unions) have traditionally done a superior job of helping people get to voting stations, giving Democratic candidates a boost of a percentage point or two that doesn't register in pre-election polls. So if you really want to know who's going to win, you have to do what the candidates themselves do: Wait until Election Day.

Data Mining

If having too little data presents a problem to the statistician, so does having too much. Imagine that 1,000 researchers are poring over 1,000 different data sets they've collected. A financial analyst is looking at the historical investment performance of different classes of stocks and bonds; a quality control expert is examining how tire-failure rates vary across different manufacturing sites and conditions; an epidemiologist is investigating how a new drug affected different health outcomes in a randomized controlled trial; a marketing manager is trying to identify the characteristics of customers who buy the most from her company's mail order catalogs; and so on. Each of the 1,000 researchers performs 10 statistical tests to determine if particular outcomes can be deemed "statistically significant." For example, the quality control expert might test whether the difference in failure rates between tires produced at one factory and another is significant.

The statistical tests indicate that 500 of the outcomes are significant. But hold on a second. The researchers collectively performed 10,000 statistical tests. Even if all of the tested outcomes were the product of chance, we would expect 5 percent of them—or 500—to be of a magnitude that was statistically significant. Do you see the problem? If you have a lot of data—and Lord knows we have a lot of data these days—then a large number of seemingly unlikely outcomes are going to occur by chance. By the same token, if you perform a lot of statistical tests on data, you will always get a good number of "statistically significant" results. We place the expression "statistically significant" in quotes to remind ourselves that being statistically significant doesn't automatically mean being significant, just as a "qualified success" is something less than a success and a "fake Picasso" is something quite a bit less than an actual Picasso.

Remember Erin Brockovich? She nailed Pacific Gas & Electric for leaching Chromium VI, a possible carcinogen, into the water supply of Hinkley, California, allegedly causing cancer and other diseases. Brockovich asserted that the incidence of disease in the area was much higher than would normally occur by chance. But Erin Brockovich wannabes should understand that an incidence greater than chance isn't enough to guarantee causality—and if the defendant's lawyers are on their toes, that's precisely the type of counterargument that the jury will hear. When you have enough data, the unexpected suddenly becomes expected. Discovering ten victims of a particular cancer in a county where only five cases would be expected may indeed defy the odds, but if you considered the distribution of that same cancer across all the counties in the United States, you'd find pockets of high incidence just by chance.

Nonetheless, researchers often scour data for differences and associations that appear significant, and then perform statistical tests to confirm their significance. The most flagrant offense is when in reporting their results, those same researchers fail to mention all of the statistical tests they conducted that *weren't* significant. Such behavior is called "data mining" (also "data snooping," "data dredging," or "hindsight") and it represents a terrible abuse of statistical science. Unfortunately, expecting statisticians to preach more judicious use of their tools is like hoping that stockbrokers will counsel their clients to trade less frequently.

What, as a consumer of data, should you do to protect yourself from data mining as well as the more general problem that chance will produce a lot of statistically significant results? First, don't forget the power of randomness. Second, pay more attention to research that was designed to test a particular hypothesis than research that uncovered something after the data were collected. Third, look for outcomes where the level of significance (the probability of such an outcome occurring by chance) is much lower than 5 percent—like less than 1 in 1000. Fourth, give more credence to a finding if there is good reason to believe it for reasons other than its statistical significance.

Lastly, always try to find the happy medium between excessive hesitancy and jumping to conclusions. "Statistics," it has been said, "means never having to say you're certain." That's a recipe for costly indecisiveness, or a stubborn refusal to acknowledge the weight of overwhelming, if not quite certain, evidence. Global warming nonbelievers come to mind. On the other hand, if you don't appreciate the power of randomness and the practice of data mining, you will reach hasty and unjustified conclusions. When it comes to interpreting the meaning of numerical data, good quantitative thinkers are neither agnostics nor suckers.

Survey Research

Special care is warranted when dealing with survey data. Information from questionnaires and interviews is particularly suscep-

tible to measurement and bias problems, since responses are often
subjective and sometimes untruthful. When, for example, surveys
ask Americans how much alcohol they drink, the answers are considerably less than what we know to be the case from sales data.
Forgetfulness (some of which, no doubt, is alcohol-induced) and
denial are probably both factors. In this section we briefly consider a few of the problems that arise in interpreting survey data.

For over two decades the Zagat Survey has asked New York
City restaurant-goers to rate their dining experiences. Restaurants
are rated on food, decor, and service, with diners voting 0, 1, 2, or
3 for each. Averages are multiplied by 10, creating a 0–30 scale,
and the results are published annually in a small maroon book. If
you don't live in an area as restaurant-dense as the Big Apple, it
may be hard to appreciate the significance of the Zagat Survey,
but it's not a stretch to say that New York City restaurant-goers
consult the Zagat Survey more religiously than Pat Robertson
consults the Bible. It's rare for New Yorkers to discuss a restaurant choice without one of them asking, "What did it get in Zagat's?" And then there's the perennial discussion about whether
the correct pronunciation is ZAG-it or zuh-GOT.

One of us lives across the street from Grimaldi's Pizzeria,
which earned a 26 rating for food in the 2002 survey. That rating
places Grimaldi's among the top 42 restaurants in New York City,
ranked by food, and in a tie with Alain Ducasse, the most expensive restaurant in the city, where dinner prix fixe menus start at
$145. Your author can confirm that Grimaldi's has excellent
pizza, but the idea that it's comparable in quality to the cuisine at
Alain Ducasse is preposterous.

Food ratings are subjective, and there are several reasons why
Grimaldi's 26 might not be equivalent to Alain Ducasse's 26. For
starters, many Grimaldi's reviewers, whether consciously or unconsciously, are presumably judging Grimaldi's pizza by compar-

ing it to other pizza and not to the kinds of dishes served at Alain Ducasse, such as "artichoke velout" and "sole, poeled whole with almonds and black truffles."

Indeed, it's likely that most Grimaldi's reviewers couldn't tell you what velout or poele means. Which hints at another consideration. Alain Ducasse's diners surely have higher standards for food than Grimaldi's diners, especially given the price differential. Not that we speak with experience, but if we were paying $280 for the "All about beef and lobster" menu at Alain Ducasse, and the food fell one iota short of the best we'd ever had, we wouldn't give it more than a 2.

A third, more subtle factor is that Alain Ducasse had been open barely a year when the 2002 Zagat Survey was published, whereas Grimaldi's had been around for ages (it was called Patsy's for most of its existence, but a lawsuit by another Patsy's forced a name change). Because Grimaldi's is well established, its patrons are much more likely to be repeat customers than are Alain Ducasse's. And repeat customers are a self-selected group of satisfied customers. Notice how this effect is likely to improve Alain Ducasse's ratings over time. The 26 rating for Alain Ducasse is an average (again, multiplied by 10), and those who gave the restaurant's food a 2 are less likely to return (and vote in subsequent surveys) than those who awarded a 3. Lesson: If you're on the fence deciding between two restaurants with comparable Zagat ratings, and one of the restaurants is a newcomer, that's the one you should pick.

Be especially wary of survey data when respondents have been asked hypothetical questions. People have a hard time predicting their own behavior, particularly in circumstances that are remote or unfamiliar. We know several people who insisted that a microwave oven was nothing more than a glorified coffee warmer—that is, until they received microwave ovens as gifts. Once they got accustomed to using a microwave several times a day, they considered it indispensable. Similarly, a golfer might tell a survey researcher that he would never spend $600 on a titanium driver, only to give in to temptation when he actually swings one in a pro shop. Longtime marketing consultants Jack Trout and Al Ries noted that in the context of a survey, the question "Would you spend $300 for an ounce of perfume?" is pretty much equivalent to the question "Are you stupid?" Yet Joy perfume, $300 price tag and all, became a smash success. Everything changed when the hypothetical became real.

Numerous companies that use consumer surveys to guide product development have learned this lesson the hard way. Ford Motor Company obtained its education in the late 1990s, when its share of the lucrative minivan market plummeted. Ford's Achilles' heel was that Chrysler and GM minivans offered four doors (not including the rear hatch) beginning with 1996 models, whereas Ford's Windstar was only available with three doors until the 1999 model year. Ford delayed the development of an optional driver's side passenger door in part because most market research indicated that prospective buyers weren't interested in the feature. For example, in 1995 (before Chrysler introduced the option of a driver's side sliding door), one major study concluded that only about 30 percent of minivan buyers considered a fourth door "desirable." Yet by late 1996, Chrysler reported that 85 to 90 percent of its minivan buyers were opting for the fourth door, a $600 option on most models, while GM claimed that 80 percent of its minivan customers were ordering dual sliding doors.[16]

Deviations from the Mean

We're almost done. But we can't very well claim to have written a chapter on statistics without covering some basic terms that you can expect to encounter time and time again. We might have begun the chapter with these terms but for our fear of sounding like a high school class. Seeing as you're still with us, perhaps we had the right strategy. Even so, we're willing to risk it all for a good cause. Here goes.

Suppose we want to analyze a data set of, say, 2,000 numbers. Already we're on thin ice. People just don't *want* to analyze that amount of data, and in our real lives, we almost never have to. But we are constantly exposed to the results of someone else's analysis of a large data set, usually in the form of an average. So if someone else is going to do all that work, the least we can do is to understand what those averages are all about.

If you followed the labor negotiations that almost led to a baseball strike in the summer of 2002, you saw two very different numbers for the "average" salary of a major league player. By one account, the average salary that season was $2.4 million. But the average was also reported as $900,000. The explanation for the apparent discrepancy is that $2.4 million represented the *mean* salary, while $900,000 was the *median*.

Let's build a simple model to try to figure out what was going on. A team has five players who respectively earn $300,000, $300,000, $900,000, $2,500,000, and $8,000,000. The mean, or arithmetic average, is calculated by adding up the salaries and dividing by the number of players. Thus, the mean is $12 million/5 = $2.4 million. The median salary is simply the one in the middle, in this case $900,000.

A couple of observations are in order. First, neither mean nor median is an inherently superior measure of "average"; the two measures provide different information. If you want to know how much our hypothetical team spends per player, you want to know the mean. By contrast, if you're interested in gauging the income of a typical player, the median is a more appropriate measure. Second, the mean is very sensitive to changes in the highest salaries, while the median is completely unaffected by the earnings of star players. If the highest salary on our model team jumped to $20,000,000—still $2 million *below* Alex Rodriguez's annual take—the mean salary would double, to $4.8 million. But the median would remain unchanged because the middle player would still take home $900,000. In fact, a widening gap between mean and median salaries is precisely what was occurring in the real world of baseball. Between 2001 and 2002 the mean salary of major league players rose 5.7 percent while the median income fell by 7.7 percent.[17] Robin Hood would have disapproved.

Now that we understand "mean" and "median," we can return to our favorite punching bag, the International Skating Union. We left the figure skating story back in chapter 2, when we described one of the purely mathematical weaknesses of the long-standing system for figure skating scoring, specifically as it applied during and before the 2002 Winter Olympics. ("Mathematical" in this context means not involving favoritism, collusion, or bribery.) After those Olympics, the call for scoring reform was widespread. One idea caught our eye because of the way it handled the nine scores of the individual judges. In this proposal the nine scores for any one skater would be condensed into a single score by first throwing out the high and low scores and then by taking the median of the remaining seven.[18] The tossing out of the high and low scores was an apparent effort to eliminate bias, but by now we realize that the median of those seven scores is exactly the same as the median of the original nine scores. With medians, what happens at the endpoints just doesn't matter.

A median is actually a special case of a percentile. Just to remind you, a percentile ranking indicates the percentage of a group that is below a given score. A student in the 90th percentile on an exam has done better than 90 percent of the kids taking the test, and so on. The median is another name for the 50th percentile. From the percentile concept we get related terms, notably "quartile," which divides a data set into four parts and is most commonly used in the context of a high school graduating class.

Just because the median is the most famous percentile doesn't necessarily make it the best. In crash tests done in connection with federal motor vehicle safety standards, the National Highway Traffic Safety Administration (NHTSA) has traditionally used the Hybrid III 50th percentile adult male dummy. The dummy is 5 feet 9 inches tall, 172 pounds, and (we're not joking) wears a "form-fitting cotton stretch short-sleeve shirt with above-the-elbow sleeves and above-the-knee length pants." But air bags that protect an average male can seriously injure and sometimes kill small adults and children. Small people are lighter and therefore more vulnerable than an average male, and if they are driving, their smaller size forces them to sit closer to the air bags. In response to some highly publicized tragedies, NHTSA is phasing in air bag requirements that will protect a Hybrid III 5th percentile adult female (4 feet 11 inches and 108 pounds), and, eventually, dummies for children down to one year in age.

There's a third measure that is often considered an average, and that's the mode. The mode is defined simply as the most common outcome. In our baseball salary example, the mode was $300,000 because two players earned that amount. If this number seems less useful than the mean and median, you're onto something. In fact, a mode doesn't really represent an average since a mode can occur anywhere within a distribution. On our baseball team, the mode was the lowest salary; that's hardly an average.

Even worse, the concept of a mode can be misleading, because "most common" needn't mean "likely." Switching to tennis for a moment, suppose you heard that the modal matchup in the finals of the U.S. Open men's singles championships was the #1 seed versus the #2 seed. Well, sure, that's the way it's supposed to be. But don't conclude that this particular outcome is likely, because it's not. During the first thirty-five years of the event (1968–2002), the top seed played the second seed in the finals on only 10 occasions, or just 28.6 percent of the time. A matchup of #1 and #2 is the mode not because it's likely per se, but because it's more likely than any other combination. (Interestingly, the #2 seed won the majority of those matches. In fact, the modal seeding of the eventual champion is #2, not #1. Second seeds won the tournament 12 times in 35 years, compared to just 8 wins for top seeds.)

Whenever you see the word "average," you should ask yourself whether that word is really supposed to be the median, mean, or mode. "The average Canadian now admits to spending C$424 ($270) per year gambling," noted *The Economist* in a 2002 article.[19] Is C$424 the median of what Canadians report gambling, or is it the mean? Probably the mean, but you can't be sure. After all, if an article said, "The average Canadian man weighs 170 pounds," we'd assume that average meant median. But even if writers did a better job of specifying their terms, the three "averages" wouldn't tell you all you need to know about a set of data, because they don't give you much information as to how the data are distributed.

When Stephen Jay Gould was given the sobering news that he had abdominal mesothclioma, a rare and incurable form of cancer, he immediately researched the disease in hopes of finding a silver lining, only to discover that the median post-diagnosis survival time was a mere eight months. Most people without Gould's savvy and curiosity would have concluded that they'd be dead within the year. But Gould realized that the median was but a single point trying to describe a more complicated distribution, and his spirits improved considerably. The distribution was in fact skewed in his favor. Though half of those diagnosed with the disease died within eight months, many lived for years afterward, especially those who were young (Gould was forty at the time) and who had the cancer identified at an early stage. Gould conjectured that the "right tail" of the distribution stretched out for many years, and time bore him out. He eventually died in 2002, twenty years after the original diagnosis. As Gould himself put it, the median was not the message.

Mercifully, we aren't going to go into detail about distributions and variations, because 1) the calculations are gnarly, and 2) the discussion could take a long time, and we're not being paid by the word. One point we will make is that variability is generally measured via a device known as a standard deviation. The general rule is simple: The higher the standard deviation, the more spread out the distribution and the more volatile the outcomes. For instance, one would expect a mutual fund concentrated on technology stocks to have higher volatility than a fund that aims to replicate a broad market index, and standard deviations confirm this. As of this writing Morningstar reports that the Fidelity Select Technology Fund has a standard deviation of 48.13 while the Vanguard Total Stock Market Index Fund has a standard deviation of 16.83. However, we should acknowledge that the term "standard deviation" is a misnomer, because there's nothing really standard about it. Furthermore, unless you know what to do with it, the number won't seem very meaningful.

More generally, standard deviations from statistical models don't necessarily coincide with deviations you might encounter in real life. The investment models used by Long Term Capital assumed that certain calamitous market events, being many standard deviations from what was considered average, weren't a concern because they would occur only once in thousands of years. But the human element behind market transactions produces a very different real-life distribution, one in which extreme outcomes are much more common than standard statistical models predict. By sticking to models with orderly distributions, Long Term Capital became a near-term disaster.

What's nice about standard deviations and other calculations of dispersion, though, is that they enable you to predict the likelihood of different ranges of outcomes. For example, suppose you wondered what the chance was that one or more of our nine .275 hitters would have done as poorly as Breuer did (batting .232) over the course of a 600-at-bat season. If you crunched the relevant numbers, you'd find that there's only an 8 percent chance of such an outcome, telling us that the very example we concocted to exhibit the power of randomness was in some sense more random than we'd ever expect.

As for the general manager who cut Breuer, we will confess some sympathy. From the front office perspective, the possibility that Breuer was an unlucky .275 hitter (as we know he was) was less likely than his being a .232 hitter with average luck. Cutting him from the team was simply playing the odds, an approach that wins in the long run but will by definition fail on many occasions along the way.

Perhaps it's fitting that we closed this chapter by returning to the baseball example. We started by questioning whether Americans could tolerate statistics, and we are reminded that we inhale more stats in the daily sports pages than R. A. Fisher did in a month of research. If we can match our innate hunger for statistics with a better appreciation of how they work, we'll be all set.

A Peace Offering for the Math Wars

This book is premised on four sets of observations. First, more and more of our world has become subject to quantification, with computers and the Internet dramatically increasing the volume of numerical information. In short, we now live in the Quantitative Information Age.

Second, every historic age places a premium on some skills and not on others. An insatiable drive to hunt down high-calorie food conferred great advantage in the Paleolithic Age, but it is less of a gift in a world teeming with Dunkin' Donuts. By contrast, the ability to make sense of statistical evidence on the pros and cons of hormone replacement therapy would be useless to a cavewoman, but nowadays such skilled quantitative thinking is more critical than ever. "Numeracy," it has been said, "is the new literacy of our age."[1]

Third, most Americans are poor quantitative thinkers. This widespread innumeracy is the father of zillions of bad decisions, and you don't need DNA testing to confirm its paternity. Numbers convey information, quantitative information. Decisions are based on information. When people are innumerate—when they do not know how to make good use of available quantitative information—they make uninformed decisions.

Fourth, there is considerable incongruence between what is taught in school mathematics and what people need to learn to become good quantitative thinkers. Much that is covered in the traditional math curriculum—absolute value, polar coordinates, long division of polynomials, for example—will vanish from most people's lives once they finish school. Well, at least until they try to help their children with math homework. And even when teaching more basic material, school mathematics typically focuses on learning abstract principles more than on identifying and applying math in real-life contexts. As an illustration, consider the following SAT I question, taken from a College Board practice exam:

> If x is 5 percent of r and r is 20 percent of s, what percent of s is x?

Everyone has to deal with percentages; that's the main reason we devoted an entire chapter to the subject. But unless you're a math teacher, you probably never face percentages expressed in such abstract terms. In any case, understanding percentages and other rudimentary math topics is no more than a beginning to good quantitative thinking. Skilled quantitative thinking hinges, above all, on having an effective approach to quantitative information. And that in turn requires developing certain habits, attitudes, and reasoning skills that are largely absent from the traditional math curriculum. In twelve years of primary and secondary school math classes, few of us are ever introduced to such practices as "only trust numbers" and "never trust numbers," nor do we hear mention of concepts like Pareto's Law or regression to the mean.

An obvious implication of these observations is that millions of people should buy and read this book. That's fine by us, and by all means spread the word, but there's an even clearer and more important implication: Quantitative education is in desperate need of reform.

But hold on a minute. It may seem imperative that the ideas

presented in this book become an integral component of school mathematics, but before any of us barges into a school board meeting, breaks down the door to a principal's office, or writes an impassioned letter to a state education department to push that view, let's take a step back and try to think prudently about the relationship between math and quantitative education.

The Math Wars

In 2000, the National Assessment of Education Progress, the federal government's ongoing survey of student achievement, determined that only 27 percent of eighth graders and 17 percent of twelfth graders were at least "proficient" in math at their grade levels.[2] A year earlier, eighth graders from thirty-eight countries participated in a repeat of the Third International Mathematics and Science Study, which was originally conducted in 1994–95. American students placed 19th in mathematics achievements, smack in the middle of the pack, and behind kids from several nations, including Bulgaria and Latvia, that are impoverished by U.S. standards. True to stereotype, the five best-performing countries—Singapore, South Korea, Taiwan, Hong Kong, and Japan—were all Asian.

Given such data, nearly everyone agrees that American schools are failing to adequately teach math. But if math education is already falling short in carrying out its existing curricular responsibilities, how can we realistically ask schools to take on the additional burden of teaching practical quantitative thinking? As stumped teachers always say, "Hmmm, that's an excellent question." The short answer is that we want to delay until high school the kind of applied quantitative education we've been discussing throughout this book, teach it primarily in separate courses, and sharpen the focus of the math curriculum onto core topics. This may sound like a blue-sky answer, and standing alone perhaps it is. So we need to take a closer look at the state of math education and the debate about its reform. Which means we have no choice but to venture into the so-called "math wars."

Now there have always been contentious disagreements about math education, as anyone old enough to remember New Math can attest. In the 1990s, however, disputes involving math teachers and professors, parents and school boards, public officials and political commentators, grew so heated that they became known as the math wars. At one point Secretary of Education Richard Riley openly called for a "cease fire."[3]

At the risk of caricaturing a complicated and multifaceted conflict, the math wars pit "progressives" against "traditionalists." Progressives want instruction that stresses problem solving while traditionalists favor a focus on skills and procedures. Progressives believe instruction should encourage students to develop their own mathematical ideas and thinking while traditionalists see teaching as disseminating information and guiding learning. Progressives maintain that the math curriculum should have more direct practical value and advocate greater use of "real-world" problems. Traditionalists push math for its own sake and defend the extensive use of notation, symbols, and abstract equations. Progressives like calculators; traditionalists hate them.

Progressives and traditionalists often seem determined not to resolve their differences. Like politicians, they spend a lot of energy attacking the opposition, with negative sound bites a common weapon. Progressives are condemned for promoting "fuzzy math" while traditionalists are denounced for backing "drill and kill" instruction. And instead of seeking common ground, progressives and traditionalists repeatedly highlight issues on which it is hard to compromise. Should students have to memorize multiplication tables? Or master the standard paper-and-pencil algorithm for long division? Should calculators be permitted in grade school? Should students learn mathematical proofs?

From this sketch of the math wars, you might figure we reside firmly in the progressive camp. After all, the aim of this book is to teach practical quantitative thinking, and at the heart of the progressive agenda is a more applied, and less theoretical, approach to math education. But as you'll see, we do not fit neatly into either camp, and in many respects we are more comfortable with traditionalist views.

Math versus Quantitative Reasoning

To say it once more, one of the most important themes of this book is that mathematical knowledge and quantitative reasoning are quite different things. Tragically, most of today's schools fail to effectively teach either. Consequently, not only are future mathematicians, engineers, and scientists sent off to college ill prepared, but a far larger group of students is left without basic quantitative resources. To boot, millions of students are turned off from math altogether, no doubt substantially lowering the number who later pursue technically demanding professions. Wherever you stand on immigration policy, it's a national embarrassment to have Silicon Valley executives lobbying for more visas for skilled workers because they can't find first-rate American engineers.

One reason it is so important to underscore the distinction between mathematical knowledge and quantitative reasoning is that being a good quantitative thinker requires very little math beyond sixth-grade levels. That sounds like good news, but it's actually a piece of bad timing that is seldom recognized. Think of some of the ways in which elementary math enters our lives: creating budgets, planning for taxes, even furnishing our homes or assessing the tax-

and-spending plans of political candidates. These issues just don't mean very much to sixth graders. Even worse, by the time these young students amass the life experience that could give the basic mathematical building blocks the relevance they deserve, those same students have gone on to trigonometry, analytic geometry, and other stops along math's traditional trail—to calculus or nowhere, whichever comes first. Precisely because the tools needed for superior quantitative thinking are so elementary, their real-world applications get lost in a system pressured to teach new material.

What all this suggests is that many of the quantitative concepts presented in this book should not be introduced until high school. More important, this process should take place in courses that are separate from the traditional math track. Many colleges have created distinct lines of courses in quantitative reasoning. It's time for high schools to follow this lead. We recognize that asking already overburdened high schools to provide different courses for mathematics and quantitative reasoning is a tall order, but we don't see another way. Imagine a "trigonometry" class that taught the definition of sine and cosine one day and then followed up that lesson with one on Pareto's Law, using the local city or town budget as a case study. It's hard to imagine that such a schizophrenic syllabus could ever work. Regardless of whether you think high schoolers should be required to learn trigonometry, the budget case study belongs in a quantitative reasoning course and not in a math class.

Another important step to improved quantitative education is to add quantitative material to other courses. "Quantitative literacy is about seeing every context through a quantitative lens," notes award-winning college math teacher Deborah Hughes Hallett. And, adds Hallett, "If quantitative literacy is the ability to identify quantitative relationships in a range of contexts, it must be taught in context. Thus, quantitative literacy is everyone's responsibility."[4] Moreover, quantitative thinking and analysis have become progressively more important in nearly all subjects at the university level, but by and large high school curricula don't reflect this development. Historical scholarship, for instance, increasingly centers on the analysis of numerical data in explaining past events and trends. Robert Fogel was awarded the Nobel Prize for economics in part for his statistical research arguing—contrary to what we were taught in school—that the development of the railroads contributed little to U.S. economic growth in the nineteenth century.[5] If high school history classes paid more attention to historical statistics and their interpretation, the classes would promote quantitative thinking and be more scholarly at the same time. That's a twofer, to put the point in quantitative terms.

The Role of Mathematics Education

Given: Schools should focus more on teaching practical quantitative reasoning. Given: Much high school math has limited immediate value for most students. But this doesn't mean that traditional math instruction should be scaled back, as many educational thinkers have long concluded. Almost a hundred years ago, William Heard Kilpatrick, a professor at Teachers College at Columbia University and an enormously influential proponent of progressive education, argued that high school algebra and geometry should be discontinued "except as an intellectual luxury." Kilpatrick maintained that such math was "harmful rather than helpful to the kind of thinking necessary for ordinary living."[6] More recently William Raspberry, the longtime *Washington Post* columnist, caused a minor ruckus when he wrote a piece in 1989 questioning the merits of algebra for all. "It's a mistake," Raspberry asserted, "to suppose that requiring the nonmathematical to take more advanced math courses will enhance their understanding and not merely exacerbate their sense of inadequacy."[7]

In a way our book agrees with Kilpatrick and Raspberry. Other than an occasional illustration in this chapter, you'd be hard-pressed to find any algebraic equations. Yes, it is possible to live a decent life without knowledge of algebra.

But the fact that you can live a reasonable life without mastering algebra is not all that powerful a reason not to learn it. By that same reckoning, it is possible to live life without knowing anything about Eric the Red, Julius Caesar, or Confucius. It is possible to live life without ever having read Shakespeare. It is possible to live life without ever having seen hieroglyphics. We could continue this list for pages and pages without writing down a demonstrable falsehood, but if you looked at the life you'd be stuck with at the end of all that, what would it be worth?

What we're getting at is that math is a cornerstone of a liberal education. The notion of a liberal education dates to ancient Greece, but the term "liberal arts" comes from the medieval curriculum of *artes liberales*. At the time there were seven liberal arts. The elementary course of study, the trivium, included grammar, logic, and rhetoric and entitled the student to a Bachelor of Arts degree. The more advanced quadrivium led to a Master of Arts degree and consisted of arithmetic, geometry, astronomy, and music, all of which were considered mathematical disciplines. (The doctoral degree was developed later, in nineteenth-century Germany.) The subjects were termed *artes liberales* because they were chosen for the intellectual training of free men and were distinguished from *artes illiberales*, vocational studies pursued for economic purposes. In short, although the meaning of liberal arts has changed over time, it has always represented an academic program that is not principally vocational.

Primary and secondary schools should do more to further practical quantitative reasoning, but they should also teach liberal arts, and that emphatically includes math. Mathematics is one of the

crowning intellectual achievements of man, and that alone justi-
fies its study. How far should students have to go in math in the
name of a liberal education? Probably through algebra, geometry,
and trigonometry, precisely because those subjects take students
beyond numerical thinking and introduce them to mathematical
notation, abstract equations, theorems, and proofs. (Ideally, we'd
require some calculus as well, but not when only 17 percent of
twelfth graders are proficient in the prerequisites.)

In truth, we suspect algebra, geometry, and trigonometry do
have practical value to a broad range of students, including those
who will never take another math course. And we don't just mean
that those subjects are useful for the SAT. Defenders of math
study often argue that the modes of thinking involved in math—
inference, deduction, proof, abstraction, symmetry—contribute
to general mental development and discipline.[8] The claim is next
to impossible to prove or disprove, but it's hard to believe there
isn't something to it.

In any case, would anyone really want a school system that
plucks out nonmathematical students and spares them from any
unpleasant experiences they might have with high school math?
By that reckoning, the kids who don't exhibit physical coordina-
tion at an early age might be exempted from gym class (to work on
their algebra, no doubt, to perpetuate a stereotype). And maybe
your authors would have been spared the humiliation of grade-
school art classes.

The remedy here is clearly worse than the disease, and it overlooks the fundamental truth that education was never meant to be
efficient. And that's not a criticism. To get efficiency, you'd truly
have to filter out kids at an early age from those disciplines at
which they are incompetent, freeing their time to concentrate on
their specialties. At least, that's the logical extension of Raspberry's argument. But we doubt that if push came to shove he'd

actually back such a system, whose implementation would feel more Communist than progressive. Society cannot try to decide in advance who in the next generation will study what, nor should it spend much time trying. Education must begin with exposure to different fields of study and different modes of thinking, and only when that exposure is complete can specialization enter the picture.

There's still another reason to introduce all students to topics beyond arithmetic. Mathematics education has a pronounced lag effect. You don't need to know algebra to understand arithmetic, but unless you've had algebra, your arithmetic skills can't possibly be brought to their fullest. Similarly, those students who have had calculus should be better at algebra than they were just a year or two earlier. The idea is that doing the work of grade $N + 2$ helps us understand the work we did in grade N. Therefore, we can justify the study of algebra as relevant even if we believe that we will never, ever have to face algebra in real life, because we're certain that we *will* have to face basic arithmetic in real life. (You can win the world pocket billiards championship without ever executing a massé shot; as a practical matter, though, any serious contender for that title has spent hundreds of hours mastering all forms of angles and english, the massé included.) In algebra classes, students must solve quadratic equations such as $x^2 - 7x + 10 = 0$, equations they will never, ever face in real life. But all that is required to solve this equation is to find two whole numbers that add to 7 and multiply to 10. Surely high school students should be able to come up with the answers 5 and 2.

In some sense this book represents the flip side of this lag effect. Although we decided not to encumber the text with complicated-looking equations, we certainly don't believe that we could have embarked on this task without the perspective offered by algebra, trigonometry, calculus, and beyond. (Yes, there is a great beyond; contrary to the perceptions of most junior high and high school students, calculus is a beginning to higher mathematics, not an end.) If you like equations, rest assured that our cutting room floor is full of them.

Unfortunately, too many math teachers lack sufficient belief in their subject to defend it on its own terms. Indeed, when adolescents ask, as they so often do, "Why do we have to learn this stuff?" many teachers shoot themselves in the foot by feigning relevance in a clumsy, self-defeating way. And textbooks often aggravate this problem. Take the textbook that "solved" an inventory problem at a manufacturing company by use of a quadratic equation. As mathematics professor Underwood Dudley, editor of the *College Mathematics Journal*, rightly noted, the idea that warehouse managers would solve quadratic equations to decide when to reorder is absurd.[9] Students may or may not be smart at mathematics, but most are smart enough to see through such legerdemain.

Less Is More

One advantage of high schools offering separate courses in quantitative reasoning is that math classes could focus on teaching math itself. When researchers with the Third International Mathematics and Science Study (TIMSS) compared math curricula and textbooks across participating countries, they found two areas in which American math education is a world leader: We cover far more topics than most other countries do and we have fatter textbooks. As far as we can tell, the only Americans who benefit from this situation are the owners of paper mills.

In the words of the TIMSS researchers, U.S. curricula and textbooks are "a mile wide and an inch deep."[10] The problem starts in the earliest grades, where American schools try to cover too many topics. The lack of focus has predictable effects, as students fail to quickly master the material. Which reminds us, there's another aspect of U.S. math curricula that ranks near the top of the world: repetition. As the TIMSS researchers put it, the American approach can be characterized as "come early, stay late."[11] The fact is, relatively few math topics need to be mastered in the first five or six grades. To achieve that mastery, it's best to study these core topics intensively, rather than letting them drag on to later grades.

Here's an illustration of how American math educators seem to favor breadth over depth. In October 2001, one of us found himself helping a fourth grader with her math homework. The girl attended an elite private school in New York City and her assignment involved converting numbers between the decimal and binary number systems. In the binary system, also known as "base 2," all values are expressed as combinations of only two digits, 0 and 1. For example, one of the homework problems called for converting the decimal (or base 10) number 502 into its binary equivalent. The answer is 111110110.

Base conversions were a staple of New Math, and Tom Lehrer rightly satirized them in a 1965 song. "Base eight is just like base ten, really—if you're missing two fingers." Although computer programmers need to understand the binary system—and, for that matter, the hexadecimal and octal systems—fourth graders do not. Nor do fifth, sixth, seventh, or eighth graders. We're talking about a peripheral topic. More to the point, it's hard to understand base conversions at much depth, and it's impossible to do them quickly, until you've mastered exponents. (The first digit in 111110110 represents 1×2^8, the second digit 1×2^7, and so on.) But as a fourth grader, this girl hadn't learned exponents yet. Consequently, her homework ended up a tedious exercise in addition and multiplication that taught her little about different number systems and diverted her from studying the more basic topics that she should have been working on. That her teacher chaired the school's math department made the situation that much worse.

Relieving the Tedium

Traditionalists are right that children should have to memorize addition, subtraction, and multiplication tables. Memorization and repetition are part of learning, and basic arithmetic operations are too important to exempt from drilling. That said, progressives are right to criticize the rote learning of formulas and procedures that often dominates school mathematics. There's no better way to alienate kids from math than to have them memorize and regurgitate algorithms. We have to make math more interesting. Students will never do more than just get by in a subject they dislike.

Regrettably, in an effort to make math more appealing, educators often make it less rigorous as well. But such a tradeoff is not inevitable. Gather a group of math aficionados and ask them why they fell in love with the subject. Not a one will say it was because math was easy. Sure, math enthusiasts found school math easier than other kids did, but people who love math seek out problems that challenge them, problems they *can't* solve. Furthermore, when you listen to accounts of why people fell in love with math, they invariably touch on the deductive logic of mathematics.

Rote learning bypasses the deductive process, the derivation of formulas and procedures. As Berkeley math professor Hung-Hsi Wu puts it, rote learning of algorithms teaches the *how* of mathematics, but not the *why*.[12] Learning the *why* makes math more interesting and rigorous, and deepens students' understanding as well.

Consider the introduction of algebra. At its core, algebra is a technique developed to determine the value of an unknown without having to resort to tedious trial and error. But how many students view algebra in that way—as something to make life *easier*? Some of the best algebra teachers deliberately expose their students to that trial and error before jumping into the *x*'s and *y*'s, in effect rolling back the clock 1,200 years to the days when Arab mathematician Muhammad al-Khwarizmi developed the primitive *al-jabr* that evolved into classical algebra. The idea is to start with something simple: If you earn $8.00 per hour, how long will it take you to earn $40? Then you give tougher numbers, first changing the earnings target to $664, then changing the hourly wage to something other than a whole number. At some level of complexity the students will either be begging for an easier way or they will develop the rudiments of algebra all on their own. Whichever position they're in, they will automatically be more receptive to what follows.

It's not hard to spot when students have been taught mathematical concepts without the proper motivation. "What is pi?" is something of a trick question for most high school students, who think they know the answer but don't. Many will reply "3.14" with remarkable self-assurance. Others will go to five decimal places and say "3.14159." But the real question doesn't relate to pi's numeric value. What is it? Why, when there are infinitely many numbers outside the world of whole numbers and fractions, does this particularly unwieldy number deserve a name of its own? The answer is that thousands of years ago, some smart folks discovered that the ratio of the circumference of a circle to its diameter is independent of the circle's size. Rather than always referring to the ratio in these cumbersome terms, mathematicians eventually gave it a name, and pi seemed as good a choice as any.

You might have noticed in the previous paragraph that we

wrote "pi" instead of using the symbol π. This raises the issue of mathematical notation, which presents a major roadblock for many students. Some educators have responded by minimizing the use of notation, but that's a detour that never gets students to their destination. Developing a facility with notation necessarily promotes the skill of abstract reasoning, and abstraction is what enables math to be powerfully applied to a wide range of contexts.[13]

As with algebra, students will better grasp and appreciate notation if they understand that it exists to make life easier, not harder. But students will still reject notation if all they're shown is that 10^8 is more efficient than its longhand counterpart of $10 \times 10 \times 10 \times 10 \times 10 \times 10 \times 10 \times 10$. To learn the language of mathematics, most students need help from ordinary language, for math expressions are much more accessible if they can be put into words. If students can't read 10^8 as "ten to the eighth power" or "ten to the eighth," not only will they balk at the notation, but they're apt to compromise by saying something like "ten eight" to themselves. The problem is that, inevitably, when you aren't saying the proper operation, you run the risk of not performing the proper operation, either. It doesn't take long for "ten eight" to mutate into "ten times eight," whereupon the student is stuck with 80 as an answer instead of 100,000,000.

Sounding out notation is only a beginning. Students should also be able to articulate what equations mean. We've seen the McKenzie brothers' approximation of $F = 2C + 30$, linking the Fahrenheit and Celsius scales. So, if it's 70 degrees Fahrenheit outside, what is the Celsius equivalent? The equation reads $70 = 2C + 30$, which quickly reduces to $2C = 40$. Would you believe that many students cannot express in words what this simple equation means? When you provide a translation ("What number, when multiplied by two, gives you 40?"), the answer $C = 20$

becomes obvious, all the more reason to be able to move back and forth between mathematical language and the English language.

Every stage of math education should encourage the habit of looking for the "easy way out." It's one thing for first graders to know that $5 + 6 = 11$, but it's far better when they understand that $5 + 6 = 5 + 5 + 1 = 10 + 1 = 11$. Similarly, what's $14 - 5$? The problem is made easier when characterized as $(14 - 4) - 1$. In both cases the easy way out speeds calculations, reduces errors, provides an anchor to the ever-present decimal system, and introduces students to the vital concept of regrouping in arithmetic. Young students who recognize these patterns are clearly proud of themselves, as they should be. On the other hand, students who must tediously count everything out before arriving at a solution are learning bad habits and boring themselves silly at the same time.

It's worth noting that the easy way out habit can also help students deal with mistakes in a more productive fashion. Consider an elementary mistake you might see in grade school: $7 \times 9 = 62$. Close, but no cigar: The actual equation reads $7 \times 9 = 63$, so we were off by 1. The good news is that in the world of whole numbers, that's the smallest mistake you can make. But recognize that 7 is an odd number, 9 is an odd number, and that the product of any two odd numbers must also be an odd number. In that sense, 62 is a really poor answer. But it provides a learning opportunity. Superior math students don't get that way because they make no mistakes. They get that way because they develop the extra perspective that enables them to see their mistakes and correct them right away.

For students trying to master their basic tables, thinking about money is often the easiest way out. It's been said that 99 percent of the arithmetic of everyday life involves money, and we don't doubt it. Hidden in that observation is how students' attention and skills can improve when problems are given a financial context.

Many high school students cannot even multiply 16×25, because the art of longhand multiplication is lost after third grade. That seems sad, and it is sad. We're talking about a very simple problem, so simple that it hardly deserves to be described as a problem, along the lines of "the toothpaste cap removal problem" or the "turn on the microwave problem." However, we aren't giving up hope quite yet. These same clueless students will find their way if you simply toss a quarter onto the table and ask how much money they'd have if they had 16 quarters. Two seconds later they will blurt out "four dollars." What's going on here? Students just aren't seeing the tie-ins between basic arithmetic and dealing with money. Perhaps "decimal" wouldn't be such a dirty word if students stopped and realized that money conventionally comes with two decimal places.

In his best-seller *Innumeracy*, John Allen Paulos suggests a greater role for puzzles in math education, and he's on the right track. Math puzzles, offered in a spirit where you're not stupid if you don't get them, provide a release from the tedium of drilling. They are to the teaching of math what "Mad Libs" are to the teaching of English: a chance to learn while goofing off. What better combination?

This process should begin in grade school, but puzzles aren't just for young kids. Consider the following puzzle of sorts, which could be used when introducing high school students to probability theory.

> Two duelists play a game of Russian roulette, sharing a six-chamber revolver loaded with a single bullet. The players alternate spinning the cylinder and firing the gun at themselves, continuing until one of them blows his brains out. What is the probability of death for the player who starts the game?[14]

What slumbering high schooler wouldn't be brought to life by this macabre puzzle? There are several approaches to solving the problem. Perhaps the easiest way out is to observe that if the first player survives the first trigger pull, then the game effectively starts anew with the second player firing first. Since the first player will survive the opening attempt 5 out of 6 times, the probability that the second player dies must be $\frac{5}{6}$ of the probability that the first player dies. By definition, the two probabilities add up to 1, so if p is the probability of the first player dying, then $p+\frac{5}{6}p=1$. Thus, the probability of death for the first player is $\frac{6}{11}$.

Included in the puzzle agenda would be mathematical curiosities. One such curiosity is the number 2,592, which is unique among four-digit numbers. It satisfies the remarkable equation $2,592 = 2^59^2$, earning a place among the "typographical errors" of famed nineteenth-century British puzzle maker Henry Dudeney. If you showed the equation to a seventh grader who was struggling in math, you might expect indifference, but you'd be surprised. Even weak math students—students who have exclaimed "I hate math" at more than one point during their brief educational careers—might say "cool" when shown a relationship such as that. Well, it *is* cool. The social irrelevance of the equation doesn't prevent it from being interesting.

The Mixed Blessing of Calculators

You might have noticed that we have barely acknowledged the existence of calculators in a mathematical education. On many occasions in this chapter someone might well ask, Why not use a calculator? Here's our reply:

When animals are placed in a zoo, they have been spared a life in the wild. The good news is that they will receive first-rate veterinary care and will not be eaten by other animals. The downside to this sheltered existence is that a normal life is rendered all but impossible. It's fair to say that cushiness is both a dream and a nightmare for all animals, humans included. The life story of Woolworth heiress Barbara Hutton *(Poor Little Rich Girl)* was the story of a curse, not a blessing. She was trapped in an ultra-rich world that hampered initiative and achievement—and she couldn't get out. Most parents aren't in a position to give their children a retailing fortune, so they do the next best thing: They give their kids calculators.

That's overstated, of course, but we wanted to get your attention. There's no doubt that calculators can be educationally useful, at least in higher grades. Calculators can provide students with immediate feedback on their work, free up time that would otherwise be spent on tedious calculations, and enable students to tackle problems that would be computationally infeasible with paper and pencil. But calculators can also be extraordinarily destructive, and it's crucial that parents and teachers understand why. Unfortunately, the National Council of Teachers of Mathematics (NCTM), the nation's largest and most influential mathematics education organization, isn't saying. Read through the 400-plus pages of the NCTM's *Principles and Standards for School Mathematics* and you will discover a veritable paean to calculator use. What you won't find is a single cautionary note.

But you don't have to spend much time in the presence of calculator-dependent youngsters to appreciate that calculators are problematic. We have seen a tenth grader from a top-rated school district unable to perform the "calculation" 51 minus 12. We have seen other students reach for their calculators to multiply 34 by 10. Even worse, when the number 340 pops up in the calculator's display, only a small percentage realize that they could have saved themselves some time by merely adding a zero. And if you stop and ask why they used a calculator for such a simple problem, you'll get the time-honored answer, "Because I wanted to be sure."

With calculators, students quickly lose sense of significant digits. Suppose you took a 12-inch strip of wood and wanted to cut it into seven equal pieces. How long would each piece be? Ask that of a calculator-wielding high schooler and you get your answer: 1.714285 inches. The fact that you can't measure beyond a tenth of an inch will be an afterthought at best. So will the notion of the width of the sawblade—the "kerf"—which renders the given answer not only silly for its six decimal places but also wrong in the real world.

Calculators obscure relationships among numbers. Let's suppose we wanted to calculate $\frac{14 \times 19}{7}$. Ninety-nine out of one hundred calculator-dependent students would multiply 14×19, getting 226, then divide by 7 to get 38. When using a calculator, it scarcely matters which operations come first, because the calculator handles them with equal facility. However, we humans don't handle all calculations with equal facility. We can drastically reduce our errors by seeking the easy way out. In this case, the numbers should be kept small by performing the division first. You divide 14 by 7, getting 2, and multiply 19 by 2 to get 38.

Calculators also deaden a student's appreciation for patterns. For example, a typical task for a geometry student would be to calculate the height, in inches, of an equilateral triangle whose sides all measure six inches. The answer, which follows from the Pythagorean theorem, is the square root of 27, or 3 times the square root of three. But calculator addicts cannot stand an abstract expression such as the square root of three. They would immediately convert the square root of 27 to 5.196. What if the triangle were 10 inches on a side? They'd start all over and get 8.66 inches. Never would they see the following pattern: You divide the side by two and multiply by the square root of three. Patterns enhance our understanding and make our mathematical lives easier, but calculators can stand in the way.

With the exception of some high-end scientific models, calculators can't do fractions. Accordingly, the rise of calculators has contributed to a de-emphasis of fractions in math education. For instance, if you look at the State of Connecticut's curricular materials, you will learn that eighth graders in the Constitution State are "not to be expected to demonstrate pencil-and-paper mastery of . . . addition and subtraction of fractions with unlike denominators, except halves and thirds or when one denominator is a factor of the other; and division with fractions or mixed numbers."[15] Translation: Eighth graders are not to be expected to know fractions. The calculation $\frac{1}{3} + \frac{1}{4} = \frac{7}{12}$ is considered beyond their scope.

Glance back at the Russian roulette puzzle. Without fractions, that basic probability-cum-algebra problem cannot be solved properly. Sure, you could convert the fractions to decimals, but not only would that be imprecise, it would also conceal the 5–6 odds that lie at the heart of the problem. Sorry, but the right answer is $\frac{6}{11}$, not 0.54545454. The Russian roulette puzzle also demonstrates what Berkeley's Hung-Hsi Wu has been telling everyone who will listen: "The understanding of fractions holds the key to understanding algebra."[16] If $7x = 3$, then $x = \frac{3}{7}$, not 0.428571. In dismantling fractions, calculators tear down algebra as well.

To get a better picture of what calculators are not good for, consider where they excel. Calculators are extremely helpful in evaluating logarithms and trigonometric functions: No one knows $\log_{10}(27)$ or $\sin(82°)$ from memory, and no one should have to. Calculators fill the void beautifully. Advanced calculators can even compute the determinants of 3×3 or 4×4 matrices, an annoying algorithmic chore that was rightfully dreaded by generations of high school and college linear algebra students. All of these concepts extend well beyond grammar school mathematics, but that's the point: Calculators belong in the hands of those who already have the elementary stuff down pat. Naïvely, the NCTM says that students in prekindergarten through grade 2 "should recognize when using a calculator is appropriate and when it is more efficient to compute mentally."[17] But kids who haven't mastered arithmetic are in no position to make that judgment. In any event, the goal is mathematical learning, not computational efficiency. If we allowed short-term efficiency to be our guide, kids would never learn to tie their own shoes because it would always be more efficient for grown-ups to do it.

What about accountants? If you've ever seen your CPA at work, you know that a calculator of some sort is never far away. But CPAs have some alibis that elementary school students lack. First, they have to make hundreds of such calculations every day, so they need to save time. Second is that they often need the "audit trail" from a calculator's printout as proof that all the relevant numbers were properly inputted. Above all, though, accountants have already paid their dues. They appreciate numbers and orders of magnitude. They can make guesses and estimates pertaining to various tax strategies long before they fine-tune the numbers. If you reverse the order—giving students calculators before they have mastered the basics—you create a gigantic problem. And that's precisely where we are today.

If calculators aren't physically jettisoned from the elementary school classroom, at least they should be harnessed to promote intuition instead of suppressing it. There are dozens of calculator-based games that encourage young students to tinker with numbers, which is precisely what they should be doing in early math classes. Technology can and should be utilized during the formative years. Interactive teaching programs not only beat the hell out of flashcards, they also help convince the budding student that the teaching methods are deserving of the twenty-first century.

Above all, we suggest two rules for all calculator use. First, students should never pick up a calculator until they know exactly what calculation they will be doing, and why. Second, students should always estimate the answer prior to performing a calculation. If their guess matches the calculator output, students can feel more confident about the answer and their intuition. If there's a discrepancy, students should seek an explanation, a process that will sharpen their estimating skills and often uncover incorrect calculations and keypunch errors.[18]

Some of what we've said in this section is harsh. But when you see high school students unable to solve third-grade problems, emotions run high. Is it frustrating? Certainly. Is it enough to make you angry? For sure. Above all, though, seeing high school students who can't add or subtract is heartbreaking. Even when students know that the calculators are to blame, their dependence endures. It's time for parents and schools to fight this national disgrace.

The Glass Is Half Full

There are reasons to be optimistic about the future of math and quantitative education in America, and there are reasons to be pessimistic. We'll start with the good news.

Americans have always had a thing for numbers. Even before computers were spitting out spreadsheets, even before baseball fans pored over box scores, Americans had a penchant for counting and measuring. In pre-Revolutionary Newport, Rhode Island, local pastor and future Yale president Ezra Stiles calculated that the city had 439 warehouses, 888 dwelling houses, 177,791 square feet of wharf surface, and 3780¼ tons of vessels in the harbor, all this alongside 77 oxen, 353 cows, and 1,601 sheep.[19] And Stiles was just a warm-up act. In the late eighteenth and early nineteenth centuries, counting and measuring turned professional. *The Picture of Philadelphia*, published in 1811, was obviously not a coffee table book displaying large glossy photographs of Independence Hall and the Liberty Bell. James Mease's work was a compilation of facts and figures about the nation's former capital, covering every- thing from the dimensions of the stage at the Chestnut Street Theatre—36' × 71', if you must know—to the number of gallons of oil burned annually in city lamps (14,355).[20] By the mid- nineteenth century, Americans were practically obsessed with quantification, a trait given headline status in historian Patricia Cline Cohen's book on the era, *A Calculating People.*[21]

As for math, our popular culture has been embracing mathe- matics as never before. Whereas in the 1939 classic *The Wizard of Oz* the scarecrow completely mangled the Pythagorean theorem even *after* he got his brain, nowadays mathematics gets more re- spect. The 1990s brought us *Good Will Hunting*, which became a smash success even though it revolved around an MIT mathe- matician. Then came *A Beautiful Mind*, which focused on Prince- ton's John Nash and the beginnings of game theory, and along the way won the Academy Award for best picture in 2002. You could argue that people came to watch Matt Damon and Russell Crowe, not to learn mathematics, but the underlying truth is that two of the hottest actors in Hollywood said yes to the thought of play- ing a mathematician. And movies weren't the only math-friendly realm of the late twentieth century. Rubik's cube, the most suc- cessful puzzle of all time, was also one of the most mathematical. And when Mattel's programmers made Barbie say "Math class is tough," the company's switchboard lit up with outrage.

Is this the picture of a society intent on rejecting mathematics? Of course not. Whatever our struggles and complaints with the subject, we also love to experience it vicariously. And most of us would like to do better with it. Popular reaction to math also questions the conventional wisdom that people dislike higher math because it isn't relevant. Andrew Wiles's proof of Fermat's Last Theorem captivated millions of people the world over. The fact that no positive integers satisfy the equation $x^n + y^n = z^n$ for

n greater than 2 is, by any reckoning, irrelevant to our everyday lives, but the *New York Times* placed Fermat on the front page the day after Wiles's proof went public. Readers appreciated the theorem's beauty even if they had no chance of understanding the mathematics that solved the problem.

The Glass Is Half Empty

Throughout this book we've come back to the point that mathematics isn't hardwired into human cells. The effects of this are visible even in first grade. Linguist Noam Chomsky has observed that even though the rudiments of language are far more complicated than the rudiments of arithmetic, young students master their native language long before they've mastered simple arithmetic.[22] That's the handicap under which mathematics education has always operated.

But it's not the only one. There's also the fabled inadequacy of American math teachers. Consider the following. Scholar Liping Ma asked Chinese and American math teachers to compute $1\frac{3}{4} \div \frac{1}{2}$ and then make up a story that required that computation.[23] Of the 23 U.S. teachers she interviewed, fewer than half performed the calculation correctly—the answer is $3\frac{1}{2}$—and only one came up with an acceptable story. All 72 Chinese teachers did the problem correctly, and 90 percent made up a satisfactory story. Here are a few of their examples:

> Yesterday I rode a bicycle from town A to town B. I spent $1\frac{3}{4}$ hours for $\frac{1}{2}$ of my journey; how much time did I take for the whole journey?

> A factory that produces machine tools now uses $1\frac{3}{4}$ tons of steel to make one machine tool, $\frac{1}{2}$ of what they used

to use. How much steel did they used to use for producing one machine tool?

Given that we want to know how much vegetable oil there is in a big bottle, but we only have a small scale. We draw ½ of the oil from the bottle, weigh it, and find that it is 1¾ kg. Can you tell me how much all the oil in the bottle originally weighed?

How extraordinary that only a single American teacher could come up with a story of this caliber. Nonetheless, American math teachers are supremely self-confident. In the 1999 round of the TIMSS research mentioned earlier, American teachers expressed a level of confidence in their preparation to teach math that was surpassed only by their colleagues from the Republic of Macedonia. An average of 87 percent of American teachers considered themselves very well prepared to teach various math topics, while an average of only 2 percent thought themselves not well prepared. Math education reform in the United States appears trapped in a Catch-22. Students can't improve until the teachers improve, but until the students improve, the teachers won't improve, because the teachers are of course former students. And the teachers can't improve until they admit they need improvement. So maybe we should import teachers from China.

And while we're at it, maybe we could import a little Asian culture, at the least a little Asian respect for academic achievement. For let's be honest: Americans may love mathematicians on the silver screen, but in the real world of the American high school, being the star of the school math team is no way to attract girls who look like Jennifer Connelly, and guys who look like Russell Crowe aren't usually turned on by girls acing AP calculus. Quite simply, social rewards in high school aren't based on academic excellence, especially in math, where accomplishment is often synonymous with nerdiness.

The Role of Government

To close this book in the spirit of unbridled optimism, here goes: We believe that the federal government could give a pivotal boost to mathematics education if its leaders would present numbers to the public in an honest, straightforward fashion. Hold your laughter as we explain.

During the presidential debates of 2000, George W. Bush derided some of Al Gore's claims as being based on "fuzzy math." Originally that term applied (either sincerely or derisively) to the efforts in California schools to teach youngsters the concepts of math rather than have them merely memorize facts about figures—as in "warm and fuzzy." Now "fuzzy" was simply derisive. It meant that someone's arithmetic was suspect.

But in a letter to the *New York Times*, esteemed mathematics professor William Thurston saw in the "fuzzy math" reference a different, even more sinister interpretation. "I gradually came to understand that by 'fuzzy math' Mr. Bush meant, 'Math is confusing and fuzzy, so ignore it.' "[24] Whether or not this was Bush's intention, Thurston had a point. For all the talk about improving our educational system (and let us not forget that the federal government contributes less than 10 percent of all education expenses), national politics has a way of shooting for the lowest common denominator when it comes to numbers and policies. The campaign of 2000 was the ultimate head spinner, because it combined genuine concern about the educational system with a host of asinine numerical arguments about everything from spending the Social Security "surplus" to the exaggerated dangers and benefits of drilling in the Arctic National Wildlife Refuge. The message conveyed by this combination appeared to be something like, "I hope this is the last generation of Americans dumb enough to believe the numbers I'm about to dole out."

Can we make politicians more accountable for their quantitative hijinks? Perhaps we need a watchdog organization along the lines of the General Accounting Office. Like the GAO, the organization would be nonpartisan. Its purpose would be to cry foul whenever a politician's numbers overstepped the bounds of common sense. An administration's programs could be graded for their intellectual honesty. Perhaps there will actually be a day when the party faithful will speak out openly on numbers that don't add up. When the libertarian CATO Institute reported on candidate Bush's claim that Arctic drilling would lessen our dependence on foreign oil, the think tank's director of natural resource studies derided the future president's arithmetic with an indictment that invoked Jeremy Bentham: "It's nonsense on stilts."[25] Whatever that meant, it had a nice ring to it.

The issue of fiddling with numbers is, of course, bipartisan. No matter what your political affiliation, it's easy to come up with examples of where the other guys have gone astray. (Or maybe where your own party has gone astray and you wish it hadn't.) Of course, if creating a truly nonpartisan watchdog organization is impossible, then we're back to where we started, with each of us going it alone. But that's not so bad: Our goal in writing this book was to help people do just that. If you find yourself more quantitatively confident than when you started, then we've all spent our time well.

NOTES

CHAPTER ONE

1. Douglas R. Hofstadter, "On Number Numbness," in *Metamagical Themas: Questing for the Essence of Mind and Pattern* (New York: Basic Books, 1985), 115–35 (originally published as a Metamagical Themas column in *Scientific American*, May 1982); John Allen Paulos, *Innumeracy: Mathematical Illiteracy and Its Consequences* (New York: Hill & Wang, 1988). Hofstadter was not the first person to use the word "innumeracy"; the *Oxford English Dictionary* credits a 1959 report on British education with that honor. For all practical purposes, however, Hofstadter invented it. In an e-mail exchange, Hofstadter told us he had not previously seen or heard the term "innumeracy" when he used it in the 1982 *Scientific American* column that introduced Paulos (and us) to the word.

2. Quoted in Patricia Cline Cohen, *A Calculating People: The Spread of Numeracy in Early America* (New York: Routledge, 1999), 132.

CHAPTER TWO

1. Quoted in Tim Kawakami, "Can't MVPlease Everyone," *Los Angeles Times*, 10 May 2000, D1.

2. See, e.g., E. M. Mandel, H. E. Rockette, C. D. Bluestone, J. L. Paradise, and R. J. Nozza, "Efficacy of Amoxicillin With and Without Decongestant-Antihistamine for Otitis Media With Effusion in Children: Results of a Double-Blind, Randomized Trial," *New England Journal of Medicine* 316 (1987): 432–37; E. I. Cantekin, E. M. Mandel, C. D. Bluestone, H. E. Rockette, J. L. Paradise, S. E. Stool, T. J. Fria, and K. D. Rogers, "Lack of Efficacy of a Decongestant-Antihistamine Combination for Otitis Media With Effusion ('Secretory' Otitis Media) in Children: Results of a Double-Blind, Randomized Trial," *New England Journal of Medicine* 308 (1983): 297–301; A. L. Olson, S. W. Klein, E. Charney,

J. B. MacWhinney Jr., T. K. McInerny, R. L. Miller, L. F. Nazarian, and D. Cunningham, "Prevention and Therapy of Serous Otitis Media by Oral Decongestant: A Double-Blind Study in Pediatric Practice," *Pediatrics* 61 (1978): 679–84. Sudafed, in case you're wondering, also appears ineffective in another of its common applications, alleviating air-travel-associated ear pain in children. It does make the little flyers drowsy, though, which can reduce the in-flight suffering of their parents. See B. J. Buchanan, J. Hoagland, and P. R. Fischer, "Pseudoephedrine and Air Travel–Associated Ear Pain in Children," *Archives of Pediatrics & Adolescent Medicine* 153 (1999): 466–68.

3. We can't remember if that's exactly what Groucho said, but we decided our research for this book wouldn't include watching every Marx Brothers movie.

4. John T. Warner and Saul Pleeter, "The Personal Discount Rate: Evidence from Military Downsizing Programs," *American Economic Review* 91 (2001): 33–53.

5. Frances Chevarley and Emily White, "Recent Trends in Breast Cancer Mortality Among White and Black US Women," *American Journal of Public Health* 87 (1997): 775–81.

6. U.S. General Accounting Office, *Prescription Drugs: Many Factors Affected FDA's Approval of Selected "Pipeline" Drugs*, GAO/HEHS-00-140 (Washington, DC: U.S. General Accounting Office, 2000).

7. Quoted in Jonathan Rauch, "Lies, Damn Lies, and Statistics," *National Journal*, 10 October 1998, 2362.

8. Ibid.

9. http://www.interpol.int/Public/Publications/SCI/default.asp (3 September 2002).

10. Christy A. Visher, "The RAND Inmate Survey: A Reanalysis," in *Criminal Careers and "Career Criminals,"* vol. II, eds. Alfred Blumstein, Jacqueline Cohen, Jeffrey A. Roth, and Christy A. Visher (Washington, DC: National Academy Press, 1986), 161–211.

11. Arthur De Vany and W. David Walls, "Uncertainty in the Movie Industry: Does Star Power Reduce the Terror at the Box Office?" Paper presented at the annual meeting of the American Economic Association, New York, January 1999.

12. J. M. Juran, "The Non-Pareto Principle: Mea-Culpa," http://www.juran.com/research/articles/SP7518.html (3 September 2002).

13. Stuart P. Beaton, Gary A. Bishop, Yi Zhang, Lowell L. Ashbaugh, Douglas R. Lawson, and Donald H. Steadman, "On-Road Vehicle Emissions: Regulations, Costs, and Benefits," *Science* 268 (1995): 991–93.

14. Jonathan H. Adler, "Leave that Car in San Francisco," *Regulation* 17 (winter 1994): 15–18.

15. "The Unkindest Cuts of All," *The Economist*, 7 April 2001, 64.

16. Leslie H. Gelb, "Misreading the Pentagon Papers," *New York Times*, 19 June 2001, A27.

17. Ingrid Gould Ellen, *Sharing America's Neighborhoods: Prospects for Stable Racial Integration* (Cambridge, MA: Harvard University Press, 2001).

18. Text as prepared for delivery, 17 May 1999, U.S. Department of the Treasury, Office of Public Affairs, RR-3152.

19. Kim A. McDonald, "Many of Mark Twain's Famed Humorous Sayings Are Found to Have Been Misattributed to Him," *Chronicle of Higher Education*, 4 September 1991, A8.

20. George Soros, *Soros on Soros: Staying Ahead of the Curve* (New York: John Wiley & Sons, 1995), 11.

21. Steven Pinker, *How the Mind Works* (New York: W. W. Norton, 1997), 343.

22. Remarks as prepared for delivery, 3 June 1998, U.S. Department of the Treasury, Office of Public Affairs, RR-2490.

23. For a discussion of Fisher and smoking, see David Salsburg, *The Lady Tasting Tea: How Statistics Revolutionized Science in the Twentieth Century* (New York: W. H. Freeman and Company, 2001).

24. http://www.brynmawr.edu/admissions/why.html (15 May 2001).

25. http://www.bcg.com/careers/interview_prep/market_sizing.asp (4 September 2002).

26. Susan Jonas and Marilyn Nissenson, *Going Going Gone: Vanishing Americana* (San Francisco: Chronicle Books, 1998).

CHAPTER THREE

1. Patricia Wen, "Size Often an Immeasurable Problem," *Boston Globe*, 23 December 1998, A1.

2. Jeffrey O. Bennett and William L. Briggs, *Using and Understanding Mathematics: A Quantitative Reasoning Approach* (Reading, MA: Addison Wesley Longman, 1999), 132.

3. Valerie Elliot, "Tesco Returns to Lbs and Oz in Anti-Metric Drive," *The Times*, 18 July 2000, 9.

4. Zach Coleman and Meeyoung Song, "Inquiry Blames Cockpit Crew for KAL Crash," *Wall Street Journal*, 6 June 2001, A22.

5. Geri Smith, "The Peso Is Falling! The Peso Is Falling!" *Business Week*, 10 July 2000, 164.

6. *Guinness World Records 2001* (London: Guinness World Records, 2000), 172.

7. *Backpacker*, March 2001, 27.

8. Jim Gorman, "Packs," *Backpacker*, March 1998.

9. Kristin Hostetter, "Ease the Burden," *Backpacker*, March 1996.

10. A. R. Behnke, B. G. Feen, and W. C. Welham, "The Specific Gravity of Healthy Men," *Journal of the American Medical Association* 118 (1942): 495–98.

11. Dale R. Wagner and Vivian H. Heyward, "Techniques of Body Composition Assessment: A Review of Laboratory and Field Methods," *Research Quarterly for Exercise and Sport* 70 (June 1999): 135.

12. Code of Federal Regulations (1–01–02 Edition), 14CFR234.2.

13. "Air Carrier Flight Delays and Customer Service," Statement of The Honorable Kenneth M. Mead, Inspector General, U.S. Department of Transportation, before the Subcommittee on Transportation, Committee on Appropriations, U.S. Senate, 25 July 2000.

14. Csaba Csere, "The Steering Column: A Gas Tax that We Could Live With," *Car and Driver*, August 2001, 13.

15. Christopher Kimball, *The Dessert Bible* (Boston: Little, Brown and Company, 2000), 14

16. Robert J. Gordon, "The Boskin Commission Report and its Aftermath," NBER Working Paper No. W7759 (Cambridge, MA: National Bureau of Economic Research, June 2000).

17. Rose Levy Beranbaum, *The Pie and Pastry Bible* (New York: Scribner, 1998), 657.

18. Beranbaum, *The Pie and Pastry Bible*, 655.

19. Jonathan Dorn, "Light and Easy Packs," *Backpacker*, June 2000, 93.

20. Jonathan Dorn, "Packs for Weekend Warriors," *Backpacker*, June 2001, 81.

21. Floyd Norris, "Dell's Share-Price Bet Cost It $1.25 Billion," *New York Times*, 3 May 2002, C1.

22. Quoted in Ann Gerhart, "One Moment in Time; For Top Athletes, Split Second Can Be an Eternity," *Washington Post*, 28 September 2000, C1.

CHAPTER FOUR

1. As cited in Hendrik Hertzberg, "Vanishing Point," *The New Yorker*, 23 July 2001.

2. John C. Bogle, *Common Sense on Mutual Funds: New Imperatives for the Intelligent Investor* (New York: Wiley, 1999), 325

3. Derrick C. Niederman, "Illusions of Grandeur," *Worth*, November 1993, 136.

4. Quoted in Robert Dreyfuss, "The Real Threat to Social Security," *The Nation*, 8 February 1999, 18.

5. Michael Crichton, "Could Tiny Machines Rule the World?" *Parade*, 24 November 2002, 6–8.

6. Mary Jo Bane and David T. Ellwood, "Slipping Into and Out of Poverty: The Dynamics of Spells," National Bureau of Economic Research, Working Paper 1199, September 1983.

CHAPTER FIVE

1. Richard Feynman, *The Character of Physical Law* (Cambridge: MIT Press, 1995), 168.

2. Neil deGrasse Tyson, "Hollywood Nights," *Natural History*, June 2002, 30.

3. Energy Information Administration, *Petroleum Supply 2001*, vol. 1 (Washington, DC: U.S. Department of Energy, June 2002), 17.

4. U.S. Bureau of the Census, "No. 828. Corn—Acreage, Production, and Value by Leading States: 1998 to 2000," *Statistical Abstract of the United States: 2001* (121st edition), Washington, DC, 2001.

5. Hosein Shapouri, James A. Duffield, and Michael S. Graboski, *Estimating the Net Energy Balance of Corn Ethanol*, Agricultural Economic Report No. 721 (Washington, DC: U.S. Department of Agriculture, 1995).

6. Housein Shapouri, James A. Duffield, and Michael Wang, *The Energy Balance of Corn Ethanol: An Update*, Agricultural Economic Report No. 813 (Washington, DC: U.S. Department of Agriculture, 2002).

7. Paul Krugman, "Outside the Box," *New York Times*, 11 July 2001, A17

8. Arnold Barnett and Alexander Wang, "Passenger-mortality Risk Estimates Provide Perspectives About Airline Safety," *Flight Safety Digest* 19 (April 2000), 1–12.

CHAPTER SIX

1. Sudhir K. Jain, Ramesh P. Singh, Vinay K. Gupta, and Amit Nagar, "Garhwal Earthquake of Oct. 20, 1991," EERI Special Earthquake Report, EERI Newsletter, vol. 26, no. 2, February 1992, accessed at http://www.nicee.org/NICEE/EQ_Reports/Uttarkashi/uttarkashi-document.htm (4 September 2002).

2. "The 10 Largest Earthquakes in the World," http://neic.usgs.gov/neis/eqlists/10maps_world.html (4 September 2002).

3. John Lamm, "2001 Chrysler PT Cruiser," *Road & Track*, May 2000, 78–83.

4. http://www.ruleof72.net/rule-of-72-einstein.asp (4 September 2002).

5. Evar D. Nering. "The Mirage of a Growing Fuel Supply," *New York Times*, 4 June 2001, A21.

6. "The 15 Largest Earthquakes in the United States," http://neic.usgs.gov/neis/eqlists/10maps_usa.html (4 September 2002).

7. Steven Lippmann, Iourii Mazour, and Hasan Shahab, "Insomnia: Therapeutic Approach," *Southern Medical Journal* 94 (2001): 866–73.

8. Robert E. Hall and Alvin Rabushka, *The Flat Tax*, 2nd ed. (Stanford, CA: Hoover Institution Press, 1995).

9. *The 2002 Annual Report of the Board of the Trustees of the Federal Old-Age and Survivors Insurance and Disability Insurance Trust Funds* (Washington, DC: Social Security Administration, 2002), 52.

CHAPTER SEVEN

1. Quoted in Herman S. Wolk and Richard P. Hallion, "FDR and Truman: Continuity and Context in the A-Bomb Decision," *Air Power Journal*, fall 1995, 2.

2. John Graunt, *Natural and Political Observation Mentioned in a following Index, and made upon the Bills of Mortality* (London, 1662), 47, accessed at http://www.ac.wwu.edu/-stephan/Graunt/bills.html (6 September 2002); R. B. Campbell, "John Graunt, John Arbuthnott, and the Human Sex Ratio," *Human Biology* 73 (2001): 205–10.

3. B. V. Gnedenko, *The Theory of Probability*, 4th ed., trans. B. D. Seckler (New York: Chelsea Publishing Company, 1962), 53.

4. George F. Will, "The Strike Zone Restored," *Washington Post*, 1 April 2002, B7.

5. John Pomfret, "In China's Countryside, 'Its a Boy!' Too Often," *Washington Post*, 19 May 2001, A1.

6. Letter from Warren E. Buffett to Berkshire Hathaway shareholders, 9 November 2001.

7. Jeremy J. Siegel, "A Terrified Economy?" *Kiplinger's Personal Finance*, December 2001, 68–70.

8. See Richard Tellis, *Once Around the Bases: Bittersweet Memories of Only One Game in the Majors* (Chicago: Triumph Books, 1998).

9. *The McLaughlin Group*, taped Friday, 26 January 2001, broadcast the weekend of January 27–29, 2001.

10. Quoted in Rick Reilly, "Order of Importance," *Sports Illustrated*, 3 December 2001, 102.

11. http://slate.msn.com/?id=32790 (5 August 2001).

12. Excerpted from *The American Heritage Dictionary of the English Language*, 3rd ed. (Boston: Houghton Mifflin Company, 1992).

13. AOPA Safety Foundation, *2001 Nall Report: General Aviation Accident Trends and Factors for 2000* (Frederick, MD: AOPA Safety Foundation, 2002).

14. National Highway Traffic Safety Administration, *Traffic Safety Facts 2000: A Compilation of Motor Vehicle Crash Data from the Fatality Analysis Reporting System and the General Estimates System* (Washington, DC: U.S. Department of Transportation, December 2001).

15. International Shark Attack File, "USA Locations with the Highest Shark Attack Activity Over the Past Decade," http://www.flmnh.ufl.edu/fish/Sharks/Statistics/statsus.htm (5 September 2002).

16. See Daniel Kahneman and Amos Tversky, "On the Study of Statistical Intuitions," *Cognition* 11 (1982): 123–41.

17. Baruch Fischhoff, Paul Slovic, and Sarah Lichtenstein, "Knowing with Certainty: The Appropriateness of Extreme Confidence," *Journal of Experimental Psychology* 3 (1977): 552–64.

18. Brad M. Barber and Terrance Odean, "Boys Will Be Boys: Gender, Overconfidence, and Common Stock Investment," *Quarterly Journal of Economics* 116 (2001): 261–92.

19. W. Casscells, A. Schoenberger, and T. Graboys, "Interpretation by Physicians of

Clinical Laboratory Results," *New England Journal of Medicine* 299 (1978): 999–1001.

20. Leda Cosmides and John Tooby, "Are Humans Good Intuitive Statisticians After All?" *Cognition* 58 (1996): 1–73.

21. See Steven Pinker, *How the Mind Works* (New York: W. W. Norton, 1997); Gerd Gigerenzer and Ulrich Hoffrage, "How to Improve Bayesian Reasoning Without Instruction: Frequency Formats," *Psychological Review*, 102 (1995): 684–704.

22. Alan M. Dershowitz, "Motive to Hit is Not a Motive to Murder," *Los Angeles Times*, 15 January 1995, M5.

23. Jon F. Merz and Jonathan P. Caulkins, "Propensity to Abuse—Propensity to Murder?" *Chance* 8 (spring 1995), 14.

24. Kevin Hayes, "Probability on Trial," http://www.ul.ie/elements/Issue5/Oj.htm (5 September 2002).

25. H. C. Bucher, M. Weinbacher, and K. Gyr, "Influence of Method of Reporting Study Results on Decision of Physicians to Prescribe Drugs to Lower Cholesterol Concentration," *British Medical Journal* 309 (1994): 761–64.

26. National Highway Traffic Safety Administration, *Traffic Safety Facts 2000*, 15.

27. D. M. Giangreco, "Casualty Projections for the U.S. Invasion of Japan, 1945–1946: Planning and Policy Implications," *Journal of Military History* 61 (1997): 521–82.

28. Institute of Medicine, *Dietary Reference Intakes for Energy, Carbohydrates, Fiber, Fat, Protein and Amino Acids* (Washington, DC: National Academies Press, 2002), S2.

CHAPTER EIGHT

1. Donald Foster, "Who Is Anonymous?" *New York*, 26 February 1996.

2. Frederick Mosteller and David L. Wallace, "Deciding Authorship," in *Statistics: A Guide to the Unknown*, 3rd ed., eds. Judith M. Tanur, Frederick Mosteller, William H. Kruskal, Erich L. Lehmann, Richard F. Link, Richard S. Pieters, and Gerald R. Rising (Belmont, CA: Duxbury Press, 1989), 115–25.

3. T. Gilovich, R. Vallone, and A. Tversky, "The Hot Hand in Basketball: On the Misperception of Random Sequences," *Cognitive Psychology* 17 (1985): 295–314.

4. Quoted in Jan P. Vandenbroucke, "Clinical Investigation in the 20th Century: The Rise of Numerical Reasoning," *The Lancet* 352 (Supplement 2)(1998): 12–16.

5. Quoted in Irvine Loudon, "Quantification and the Quest for Medical Certainty," (book reviews), *The Lancet* 346 (1995): 1692–93.

6. Gerard E. Dallal, "The Regression Effect / The Regression Fallacy," http://www.tufts.edu/~gdallal/regeff.htm (17 April 2002).

7. Francis Galton, "Typical Laws of Heredity," *Nature* 15 (1877): 492–95; 512–14; 532–33.

8. Myra L. Samuels, "Statistical Reversion toward the Mean: More Universal than Regression toward the Mean." *American Statistician* 45 (1991): 344–46.

9. Charles T. Clotfelter and Jacob L. Vigdor, "Retaking the SAT," unpublished paper.

10. Quoted in Tony Schwartz, "The Test Under Stress," *New York Times Magazine*, 10 January 1999, 35.

11. Alexander Wolff, "Unraveling the Jinx," http://sportsillustrated.cnn.com/inside_game/alexander_wolff/news/2002/01/15/wolff_viewpoint/ (19 April 2002).

12. Peter L. Bernstein, *Against the Gods: The Remarkable Story of Risk* (New York: John Wiley & Sons, 1996) 6.

13. Quoted in Brian S. Everitt and Andrew Pickles, *Statistical Aspects of the Design and Analysis of Clinical Trials* (London: Imperial College Press, 2000), 4.

14. John A. Barker, "Lind and Limeys, Part 1: A Brief Early History of Scurvy and the Search for Its Cure in the 18th Century," *Journal of Biological Education* 26 (spring 1992): 45–53; John A. Barker, "Lind and Limeys Part 2. On Limes, Lemons and Guinea-Pigs: Scurvy in the 19th Century," *Journal of Biological Education* 26 (summer 1992): 123–29.

15. Richard A. Deyo and James N. Weinstein, "Low Back Pain," *New England Journal of Medicine* 344 (2001): 363–70.

16. Drew Winter, "Door Wars," *Ward's Auto World*, November 1996, 50–52.

17. "Earning Power: Baseball's Average Salary Climbs to Nearly $2.4 Million," http://sportsillustrated.cnn.com/baseball/news/2002/04/03/salaries_ap/ (7 September 2002).

18. E. M. Swift, "Figuring It Out," *Sports Illustrated*, 13 May 2002, 19.

19. "Hard Luck Story," *The Economist*, 14 September 2002, 37.

CHAPTER NINE

1. Lynn Arthur Steen, "Preface: The New Literacy," in *Why Numbers Count*, ed. Lynn Arthur Steen (New York: College Entrance Examination Board, 1997), xv.

2. U.S. Department of Education, Office of Educational Research and Improvement, National Center for Education Statistics, *The Nation's Report Card: Mathematics 2000*, NCES 2001–517, by J. S. Braswell, A. D. Lutkus, W. S. Grigg, S. L. Santapau, B. S.-H. Tay-Lim, and M. S. Johnson, Washington, DC, 2001.

3. Remarks of Richard W. Riley, U.S. Secretary of Education, "The State of Mathematics Education: Building a Strong Foundation for the 21st Century," Conference of the American Mathematical Association and the Mathematical Association of America, 8 January 1998.

4. Deborah Hughes-Hallett, "Achieving Numeracy: The Challenge of Implemen-

tation" in *Mathematics and Democracy: The Case for Quantitative Literacy*, ed. Lynn Arthur Steen (Princeton: The Woodrow Wilson National Fellowship Foundation, 2001), 94–95.

5. See Robert William Fogel, *Railroads and American Economic Growth: Essays in Econometric History* (Baltimore: Johns Hopkins Press, 1964).

6. Quoted in David Klein, "A Brief History of American K-12 Mathematics Education in the 20th Century," http://www.csun.edu/~vcmth00m/AHistory.html (18 September 2002).

7. William Raspberry, "Math Isn't for Everyone," *Washington Post*, 15 March 1989, A23.

8. See, for example, Charles Van Doren, "Mathematics," in *The Paideia Program*, ed. Mortimer J. Adler (New York: Macmillan, 1984), 71–85.

9. Underwood Dudley, "Is Mathematics Necessary?" *College Mathematics Journal* 28, no. 5 (November 1997): 360–64.

10. William H. Schmidt, Curtis C. McKnight, and Senta A. Raizen, *A Splintered Vision: An Investigation of U.S. Science and Math Education* (New York: Kluwer Academic Publishers, 2002), 62, 122.

11. Ibid., 20.

12. Hung-Hsi Wu, "The Mathematician and the Mathematics Education Reform," *Notices of the American Mathematical Society* 43 (December 1996), 1531–37.

13. Gail Burrill, Susan Ganter, Daniel L. Goroff, Frederick P. Greenleaf, W. Norton Grubb, Jerry Johnson, Shirley M. Malcom, Veronica Meeks, Judith Moran, Janet P. Ray, C. J. Shroll, Edward A. Silver, Lynn Arthur Steen, Jessica Utts, and Dorothy Wallace, "The Case for Quantitative Literacy" in *Mathematics and Democracy: The Case for Quantitative Literacy*, ed. Lynn Arthur Steen (Princeton: The Woodrow Wilson National Fellowship Foundation, 2001), 1–22.

14. Borrowed from Gunnar Blom, *Probability and Statistics: Theory and Applications* (New York: Springer-Verlag, 1989), 32.

15. *Connecticut Mastery Test Third Generation Mathematics Handbook* (Hartford: Connecticut State Board of Education, 2001), 6.

16. Hung-Hsi Wu, "On the Learning of Algebra," http://math.berkeley.edu/~wu/algebra1.pdf (10 September 2002).

17. *Principles and Standards for School Mathematics* (Reston, VA: National Council of Teachers of Mathematics, 2000), 77.

18. See David Maister, "How to Avoid Getting Lost in the Numbers," Harvard Business School note 9-682-010, 30 September 1985.

19. Patricia Cline Cohen, *A Calculating People: The Spread of Numeracy in Early America* (New York: Routledge, 1999), 111.

20. http://ascc.artsci.washington.edu/drama/mease.html (4 September 2002); Cohen, *A Calculating People*, 154.

21. Cohen, *A Calculating People*, 4.

22. "Although children receive no instruction in learning their native language, they are able to fully master it in less than five years. This is all the more confusing as language is much more computationally complex than, say, simple arithmetic, which often takes years to master." Quoted in "Obituary: Kenneth Hale," *The Economist*, 3 November 2001, 89.

23. Liping Ma, *Knowing and Teaching Elementary Mathematics* (Mahway, NJ: Lawrence Erlbaum Associates, 1999).

24. William Thurston, "Fuzzy Math," letter to the editor, *New York Times*, 6 October 2002, A30.

25. Jerry Taylor, "Bush's Energy Babble," 30 September 2000, http://www.cato.org/dailys/09-30-00.html (10 September 2002).

INDEX

ABOUT THE AUTHORS

Derrick Niederman is a financial writer and a senior contributing editor at *Worth* magazine. His previous books include *The Inner Game of Investing*, *A Killing on Wall Street*, and several volumes of math puzzles. He received his B.A. from Yale and Ph.D. from MIT, both in mathematics. He lives in Needham, Massachusetts.

David Boyum is an independent consultant whose clients have included major corporations, government agencies, law firms, and private investors. He received his A.B. in applied mathematics and Ph.D. in public policy from Harvard. He lives in New York City with his wife and two children.